E-volunteer
Revolution

E-volunteer Revolution

Matching Missions and Skillsets to Unlock Digital Altruism

D.L. Frugé

Red Badge

Copyright © 2017 by David L. Frugé. All rights reserved. Printed in the United States of America. Except as permitted under the United States Copyright Act of 1976, no part of this publication may be reproduced or distributed in any form or by any means, or stored in a database or retrieval system, without prior written permission of the publisher.

ISBN-13: 978-1985827530

ISBN: 1985827530

This publication is designed to provide accurate and authoritative information in regard to the subject matter covered. It is sold with the understanding that neither the author nor the publisher is engaged in rendering legal, accounting, or other professional service. If legal advice or other expert assistance is required, the services of a competent professional person should be sought.

> —From a Declaration of Principles Jointly Adopted by a Committee of the American Bar Association and a Committee of Publishers and Associations.

Red Badge books are available at a special quantity discount to use for nonprofit and corporate training programs. To contact a representative, visit the Contact Us page at www.RedBadge.org.

For the men and women who opened my eyes to the lack of good jobs and opportunity, and to my wife Katie who joined me in putting our wants aside in order to meet the needs of others.

Contents

Author's Note xiii
Introduction 1
Chapter 1. Digital Altruism 5
Chapter 2. Big Shifts in Volunteering 12
Chapter 3. Surprising Stats 28
Chapter 4. Big Secrets in Volunteering 35
Chapter 5. Bad News: Inefficiency 45
Chapter 6. Good News: Potential 51
Chapter 7. Volunteers Are Healthier 61
Chapter 8. Mission and Passion 66
Chapter 9. Pitfalls and Obstacles 72
Chapter 10. Just for Nonprofits and Small Businesses 79
Chapter 11. Just for E-volunteers 90
Appendix A. The Best Apps and Online Tools 107
Appendix B. E-volunteer Resources 123
Appendix C. The Best Social Media Platforms 127
Appendix D. E-volunteering Case Study: Hurricane Harvey 132
About the Author 139
About Red Badge 141

E-volunteer
Revolution

Author's Note

I couldn't believe volunteering hadn't been brought into the digital age! Volunteering and altruism are foundational parts of the human experience—and perhaps the last parts to be brought into the 21st Century and the digital age.

I'm excited to write this book because I love people and I'll always side with the underdog! Just like you, I cannot stand bullies, but I also believe in helping the bullies after I've first helped the oppressed. I care a lot about relationships, and I've witnessed how in America our relationships have suffered even though we're busier than ever. It's my hope that you'll rethink what you're passionate about. Less than 5% of people can earn a living doing what they're passionate about—and living without passion leaves a void that all our busyness can never fill.

In my own life, I regained purpose and joy when I started volunteering a little every week to help young fathers find their first steady incomes. It broke my heart to see a good person be unable to even get a chance. That's when I determined to do everything I could to give people a fair chance at the life so many of us take for granted. For me this meant helping get people's lives back on track via assistance from a nonprofit, and then keeping people's lives on track by helping them get a good job or promotion at a local or large business.

It takes a lot of effort to help with nonprofits and employment. This led me to discover the untapped potential of virtual volunteering. I couldn't believe volunteering hadn't been brought into the digital age! Volunteering and altruism are foundational parts of the human experience—and perhaps the last parts to be brought into the 21st Century and the digital age. I'm excited for it because it's something both the benefactors and the beneficiaries need more than they realize. Welcome to the e-volunteer revolution!

Introduction

Sometimes—in fact many times—the great causes and movements the world needs most have no one willing to pay for them except the volunteers paying with their time and dedication.

Volunteering hasn't changed since the 1700s! That's over 300 years ago! That doesn't mean volunteering hasn't been effective. For 300 years, volunteers have been changing the world by dedicating themselves to causes they care about. Your time is precious, and you have an incredible opportunity to make the world a better place. When you're part of updating volunteering for the digital age, you'll be part of the movement to speed up and spread improvements in your neighborhood and across the world!

Volunteering has always been about people and individuals giving away their time for free because they want to make some aspect of society or the world better. Do you realize all the good volunteers do? Volunteers work with their hands to help at shelters, to package

E-volunteer Revolution

food, to spread awareness, to campaign for improvements in the legal system, to fight for equality, to preserve the earth for future generations, to give families hope, to give mothers hope, to give kids hope, and to do many other incredible things that individuals and groups are passionate about.

Humans have always been volunteering—even before volunteering became a recognized service 300 years ago. People have always give their time, their years, and even their lives for causes they care about. Sometimes—in fact many times—the great causes and movements the world needs most have no one willing to pay for them except the volunteers paying with their time and dedication.

Volunteering will continue as long as any part of the world needs to become a better place, whether that's in your neighborhood, across the ocean, in your kids' school, or in your own family. We will always be giving our time to those in need because that is part of being human. However, we only have so much time, and we have to be wise about how we use our time. Many people have looked back with regret for lost years after giving time and money to causes and movements that either were less than effective or became part of the problem rather than part of the solution.

There is good news! Your time and your passions are no longer confined to the past 300 years of traditional volunteering. Technology has already brought over 30,000,000 global businesses into the digital age. Volunteering has lagged behind. But that is over because millions of volunteers around the world have been bringing volunteering into the 21st Century and making volunteering digital from start to finish.

I am personally energetic about this because my

Introduction

passion is to see the world become a better place by making good jobs, incomes, economic freedom, and self-determination available to all people. I know that volunteers play a crucial role in helping individuals, families, and groups get back on track. I also know that having your life together is just the first step in being able to be part of the socio-economic system on your own terms and goals instead of trapped in others' low or misguided desires for you.

I have seen volunteering transform families and neighborhoods. I have also seen volunteer organizations experience a transformation when they realize having computers at their nonprofit is not enough. It's simple. Stop using your computers just for email and Microsoft Office. Use the Internet. Use the Internet to find e-volunteers. Then free those virtual volunteers to help you do anything that doesn't 100% have to be done in the nonprofit's actual office. Next, free those passionate virtual volunteers to show you everything else that you can start doing digitally to make a bigger impact! You'll be surprised!

Yes, you will be surprised what we virtual volunteers can do if you'll just ask us for what you need or describe what you want to do. Even tell us what's causing you pain—we care about your mission to make the world better, and we can do a lot! The biggest companies are all digital. Their employees—like all humans—like to volunteer for things they care about. Give them the opportunity! Let's all work together to bring volunteering completely into the digital age and the 21st Century!

In this book I will cover the revolutionary changes happening right now in volunteering. I am excited to reveal some surprising statistics and secrets about e-volunteering and traditional volunteering. We'll quickly cover the inefficiencies in traditional volunteering in order to get into the unlimited potential of e-volunteering. We'll look at the health of volunteers and the importance of mission and passion in life. After covering the pitfalls and obstacles facing volunteering, we'll conclude with chapters just for organizations and also just for e-volunteers.

At the end of this book I hope I will have convinced you to virtually volunteer not just for nonprofits but for local businesses and socially-minded startups too! You'll see in this book that you're already volunteering consciously or unconsciously for big businesses in social media and entertainment. You might as well virtually volunteer on your own terms—not theirs—for causes and companies whose passions align with yours.

Most importantly, we all know about the billion dollar companies that began in garages and basements, but it's my dream that the 21st Century's best nonprofits and socially-minded companies won't be birthed out of a garage or a basement but instead will be created, maintained, and incubated in soup kitchens, in speech therapy rooms, and in hospital wards. It can and will happen because virtual volunteering can be done from anywhere. With the right systems, processes, culture, and passion in place, e-volunteers can create anything. Let's work together to update volunteering for the 21st Century as soon as possible. Those in need and volunteers are counting on us. Best of all, e-volunteering is healthy, exciting, and rewarding.

Chapter One
Digital Altruism

How is it possible that the Internet has changed everything about the world except volunteering?

It's unbelievable that volunteering hasn't changed since the 1700s! The stagnation is understandable up until 1990. However, how is it possible that the Internet has changed everything about the world except volunteering? The Internet has changed how businesses run, how businesses get started from nothing, how people communicate, and how entire countries are judged. Volunteers are passionate about making the world better, and to make the world better, volunteers and nonprofit organizations need to use all the same digital tools and remote work used by companies.

In the 21st Century, it does not make sense for a nonprofit food pantry to ask a software developer who makes $100 per hour coding Java, JavaScript, and Perl to come into your food pantry and do an unprofessional job painting the walls the best she or he can for the equivalent of $5-7 per hour at replacement value. It's time for that nonprofit food pantry to get help from an e-volunteer to fix the bugs in their website or consult on the best path forward to build up the online portion of the nonprofit.

People who work in their professions making $20, $30, $50, and $100 per hour want the option to volunteer with their own skills to make the world a better place. Let's not waste their time. Doctors and dentists are currently not volunteering to go to impoverished communities and countries to paint the walls of the food pantries and clinics just to fly home before the paint dries. No. Doctors and dentists are going overseas to volunteer their time and medical experience to heal the sick, correct blindness, and save lives. Doctors, dentists, and nurses are being asked to volunteer their $30, $60, $100, $200, and $500 per hour skills to make the world a better place. They aren't the only ones who have business skills that translate to volunteering.

We must start allowing everyone to volunteer their skills to make the world a better place! The easiest way to do this is referred to interchangeably as e-volunteering, virtual volunteering, online volunteering, and remote volunteering. While there are some highly valuable skills that must be done in person—such as those in the medical field—many businesses already rely on a plethora of jobs that are done remotely on a laptop from the employee's home or favorite coffee shop. It's time for nonprofits to seek virtual volunteers to complete virtual tasks. It's time for nonprofits to dream big and ask virtual volunteers what a new phone app would cost to develop and whether the software development could be completed by a team of virtual volunteers. Yes! This is the e-volunteer revolution.

Digital Altruism

Request Skills Not Just Financial Donations

It's incredible when you realize that nonprofit organizations do not need people's money as much as they need people's skills. Nonprofits are overly reliant on a small pool of donors who often do not have the specific skill needed but do have money to donate. The problem that nonprofits run into is that they have to pay for most of the things they need, and they go without the things they can't afford. Nonprofits need e-volunteers' skills as much as they need donors' money. When appropriate, E-volunteering needs to replace traditional volunteering.

Nonprofits need to stop seeking out volunteers' time to do menial tasks and instead seek to empower volunteers to donate their specialized skills, advice, and network. Volunteering in the knowledge economy should look different. It currently doesn't, but it's an easy fix to make.

Nonprofits need to look at everything they are doing and everything they want to do. First, examine current expenses, payroll, and projects. Are you currently tracking which areas of the nonprofit are being supplemented with on-location volunteers? If not, then try to set aside time to understand how volunteers at the nonprofit are plugged into a system and which ones are just being thrown randomly into tasks.

If you know what are the parts running your nonprofit's "car" then you can figure out how to make some of those parts virtual tasks which your current volunteers—and new recruits too—can help perform from home as e-volunteers. (Tip: having some opportunities for e-volunteer can bring in a lot more new volunteers and also help with retention.) There are a lot

E-volunteer Revolution

of parts that run the nonprofit's "car" ranging from phone receptionists to computer programs for accounting, CRM, and documentation. Figure out which of these parts can be outsourced to e-volunteers. Much of your fundraising, awareness, and graphic design can be outsourced to virtual volunteers excited to help and who share your passion. There is no geographical limit to e-volunteering.

Take it a step further and dream big about things you wish your nonprofit had the money to do. People running nonprofits have an insider's perspective on what needs to be done. So if you've got that insider's knowledge built up over years of helping people, share what you need. It's not impossible with e-volunteers. It's more than likely that virtual volunteers know what type of software is necessary to create that project for free. In the tech world, we call this open sourcing, and programmers are always looking for projects to do to sharpen their skills or develop new skills. You would be surprised at how many e-volunteers would rather help a nonprofit create a new phone app to make your nonprofit better—and maybe the new app will be used by similar nonprofits everywhere! (Hint: people are attracted to purpose. E-volunteering removes a lot of the obstacles.)

Make sure you ask for skills not just money because sometimes skills are all that people have to give. Someone might be in student debt or credit card debt and feel unable to contribute money to a cause she or he loves. Eager volunteers might not have time to drive an hour roundtrip just to volunteer a few hours, and they might not be able to schedule their time in advance like you need them to. However, if you give them a virtual assignment like making a new logo, designing a new poster, or auditing your social media then they'll be excited to do it and you'll get a $100-400 task for free. It's a win-win! It's the e-volunteer revolution.

Volunteer at What You're Worth

Don't settle for bagging old clothes or separating out rotten tomatoes—unless you want to. Consider what you do to make money for yourself and/or your family. Whether you make $10, $20, or $80 per hour, go find a nonprofit that will allow you to do what you're good at so that you can make the maximum impact. Remember to let nonprofits know if you have a side-hustle or digital-gigs that you do too. There are many great people who do side-projects for $50-100 per hour, and their next side-project could be a virtual volunteer project to make the world a better place.

If you've got a skill or are developing a skill that is highly sought after in the knowledge economy, then make sure you don't miss out on using that skill to help others well. (Tip: a great time for bagging old clothes or separating out rotten tomatoes is when you have time to be at the nonprofit, can make sure that you have a joyful attitude, and want to be able to relax and hopefully complete the tasks with other people involved to benefit your social well-being!)

Consider Volunteering for Businesses

This sounds crazy at first, but a lot of people are already volunteering for businesses. When you get your oil changed, cancel a service, or go on a vacation, you receive requests for reviews and surveys. Sometimes the requests come from the service provider and sometimes from the broker who connected you with the service. When you fill out their reviews or surveys, you are virtually volunteering your time for their businesses. Why do you do that? Will you stop? Most people provide reviews and fill out surveys because they appreciate the business enough to want to help it improve or encourage it to continue providing a great product.

If you're going to volunteer your time for a business, then at least volunteer for what you are worth for a company you really love. If you feel guilty because you stopped going to a local tailor even though you love local businesses, consider whether you can do any e-volunteer tasks to help. If another parent at school starts a business to get more greenery and plants inside apartments—and you love this—then consider whether you can e-volunteer in any way to help. It's time to start making a bigger impact with the same amount of time—because some of those surveys take 30 minutes, and we just want to help—and it's easy to answer questions. It's also easy to do a skill you've perfected for a cause you love—nonprofit or business.

There are a lot of ways that people are volunteering unaware for businesses besides responding to reviews and surveys. Referrals, suggestions, comment boxes, and word of mouth advertising are a few. We volunteer for individuals too when we help them get a job through our network, help them with a task at work, and send business their way. Humans enjoy helping other people and nonprofits—and even businesses—because it makes us happy and feel more secure. It's my hope that volunteers will maximize their efforts by

Digital Altruism

virtually volunteering their most effective skills. It's a further hope of mine that socially-conscious startups and local businesses will join nonprofits to ask e-volunteers for help to make the world a better place.

Chapter Two
Big Shifts in Volunteering

Ben Franklin volunteered at what he was best at doing, and he gave his time freely to wherever he could make the biggest impact on people's lives.

The roots of volunteering in America predate the Declaration of Independence. In the 1730s Ben Franklin was the most famous of the early American volunteers. He encouraged everyone to volunteer. He practiced what he preached and volunteered more than anyone too! But he did not waste his time. For free, he started volunteer fire departments such as the volunteer fire company of Philadelphia. Ben Franklin founded militias, libraries, agricultural schools, hospitals, mutual insurance companies, and intellectual societies.

Ben Franklin added more value to people's lives than anyone else in America. He didn't volunteer and improve lives by signing up to paint the walls of the volunteer fire department. No. He founded the volunteer fire department and then went on to volunteer his specific skills. Ben Franklin allowed Philadelphia's

Big Shifts in Volunteering

painters to volunteer to paint the fire department—and they loved contributing. Ben Franklin volunteered at what he was best at doing, and he gave his time freely to wherever he could make the biggest impact on people's lives.

Ben Franklin's personal volunteering and encouragement to others to volunteer was part of the American ethos in the 1700s. The American Revolutionaries were not a mercenary army fighting only for themselves against the British. The American Revolutionaries were fighting and dying for their loved ones. They were all facing death. All the American leaders were going to be hung or executed by firing squad if the British won. It was not a power grab. It was not a sure thing that the Americans would win. The American Revolutionaries were volunteers—and the war they won allowed their ethos of volunteering to continue.

Look again at all Ben Franklin did as America's biggest supporter of volunteering as the means to improve society. Ben Franklin might have swept his own home's floor, but when it came to sweeping the floors of charities, public hospitals, and public libraries, he left that to others—but he was always ready to grab a broom if a demonstration of work ethic was in order. He cared about people, and he knew that institutions like intellectual societies make a long term impact.

These same altruistic achievements can, and are, being done today. Intellectual societies are forming under different names and using different mediums. Some are large and some are small. Many people are imitating Ben Franklin to create nonprofits that can make a big impact on people's lives. But look at the difference between the nonprofits and socially-conscious companies which are making a big impact versus those that are not. Have you seen the difference? Why do some succeed and others fail to make a difference?

The companies and nonprofits struggling to make a big impact might share Ben Franklin's volunteer ethos, but they are also operating like it's still the 1700s. They have traded their horse and carriage for a car, their letters for a phone, their ships for plane tickets, and their quill for a laptop, but they're missing what made Ben Franklin Ben Franklin. The right tools must also have the right use!

Ben Franklin sought to make the biggest difference possible. He would publish—even under a pseudonym if necessary. He would take advantage of everything possible to do more. In many ways, Ben Franklin's American descendants aren't on farms any more. They are not on factory floors or making deliveries. Ben Franklin's descendants are in the tech sector, and they're already virtually volunteering to make the world a better place. Virtual volunteers already made the world's top open source website platform and also the world's top open source online encyclopedia.

The nonprofits and small businesses willing to share their passions and needs with virtual volunteers will find that not just Americans, but people around the world, enjoy helping others make the world better. Ben Franklin's legacy is passion, hard work, helping to organize the effort, thinking outside the box, and seeing things through to completion.

In this chapter we'll continue to look at the history of volunteering in America, the digital shifts that are occurring, and who will be the winners and losers in the age of digital altruism.

Big Shifts in Volunteering

America Is a Volunteer Nation

America has volunteering as part of its ethos. It's easy to see why. America as it is in the 21st Century exists because people in England suffering from different forms of injustices and economic restraints joined together to cross the ocean in leaky and disease-filled ships. The voyagers had to support one another in the ships and make hard decisions in close-confines to be for or be against fellow voyagers.

Arrival in North America meant the beginning of new hardships. The pilgrims and the colonists had to help each other just to survive. They had to first build a pubic house for food, assembly, and worship. Then they had to build their homes one at a time, trusting others to help after their own home was finished. Perhaps more than anything, they had to trust each other to take care of their children should the parents die in those harsh conditions. The spirit of volunteering and extraordinary altruism was growing in the Thirteen Colonies. Where there was no volunteering there was no survival.

Then after surviving in the Thirteen Colonies, the Americans had to work together to thrive. They wanted more than what they left behind in England, and they got it. The average Americans were taller and more nourished than their countrymen and women back home in England. The Americans began to become pioneers not just pilgrims. The Americans survived and thrived in part because they created the first country that allowed prosperity based on skill and merit instead of on religious affiliation. Americans created what is commonly called a "laboratory" in which to achieve a certain cause or goal people of different religious convictions put away all things except the goal and the scientific processes and tasks to achieve the goal. (Hint: this is one of the central keys to success—putting aside differences that do not achieve the immediate goal at hand.)

This spirit of volunteering to help one another despite differing core convictions has been passed down in America through the generations and commonly exported as democracy—although perhaps this short history will show why American-styled democracy is just the best label available for America's passion to export and spread the freedom to do what's best for your own family and to volunteer to help other families.

These collaboration efforts and group—almost cult-like—dedications to goals have been titled "startups" by the business world. Some startups are socially conscious endeavors that have supplied millions of shoes to impoverished people—but only when someone in America buys a shoe. Nevertheless, although a for-profit start up, it did immeasurably more good than harm, and the amount of shoes provided was staggeringly more in impact and awareness than any nonprofit had achieved in that area to date.

However, the point is that an incredible amount of volunteering, collaboration, and loyalty has been fully embraced in the business community in America and around the world. It's time for nonprofits to take true advantage of America and the world's DNA for volunteering. It's time to take volunteering online! The e-volunteer revolution is underway!

Digital or Die

Volunteering needs you, and volunteering needs to go virtual. If volunteering doesn't follow the global

Big Shifts in Volunteering

business community and incorporate digital tasks and virtual workers, then volunteering will continue to decrease. Without the power, efficiency, and network effects of virtual volunteering, traditional volunteering's role in the world will be diminished to a few huge global nonprofits and tiny-ineffective family-run nonprofits. The middle will disappear, and the middle is where all the ingenuity, creativeness, and future growth occurs. The middle-size nonprofit gives birth to other nonprofits and passionate people. The middle-size nonprofit organization is like the middle class in America—both nonprofits and the middle class, without embracing technology, will decrease.

Virtual volunteering must be embraced because it unlocks potential for unimaginable growth, global awareness, and ripple effects of positive impact for generations. Companies could not function without relying on online skills, relying on online applications, outsourcing online tasks, and hiring remote workers. The only reason nonprofits can function and crawl along without many or all of these things is because nonprofits have no choice since they often lack the budget and resources to do these things.

Most nonprofits don't sell products so they often rely on donations. While we want nonprofits to start asking for skill donations instead of only financial donations, the current status is that over 99% of nonprofit requests are for money not skills. This will change when nonprofit directors see the benefits of asking for high value tech skills donated by e-volunteers. However, the current situation is that nonprofit directors are most burdened with fundraising and preaching the importance of their cause. Nonprofit directors have a great insufficiency of knowledge about how to incorporate technology and virtual workers into their nonprofit. Sure they know a little, and they could give a local dry cleaning business a run for their money regarding which entity is more tech savvy. But there is much more available, and there is a huge need to onboard nonprofits into the digital age and virtual volunteer marketplace.

Innovation Creates 10x's Multipliers

Henry Ford was a disrupter who innovated and made the world a better place. Henry Ford built the assembly line which created efficiency and improved the lives of workers—and the general public. Thanks to Henry Ford, cars were no longer available only to the rich, but it's what happened for the workers that is particularly amazing.

Now it is well known that Henry Ford did not have a sparkling record with his workers' housing or the unions, but what goes unnoticed is a modern miracle of efficiency. The efficiency of the assembly line allowed workers to go from making fifty cents to making five dollars. That is a ten-fold increase credited to the interworking of technology, efficiency, and risk. Even better is that manufacturing cars is not the only area of the world that a ten times multiplier can be created.

Martin Luther King Jr's speeches are legendary and part of the American culture. If they were not written down or recorded, then we wouldn't know about them. While this has little to do with a ten times multiplier, it does deal with technology and a public figure who has had an outsized impact on me and many people around the world. Many things you do can have huge ripple effects if the right people get involved. Many huge businesses originally came into existence because a visionary person came across a great product, a hamburger, a chicken sandwich, an Italian coffee shop, or a personal computer.

Technology, risk, and collaboration among nonprofits, virtual volunteers, and, yes, social entrepreneurs, can make the world better for everyone. Can you imagine making ten times more money? Can you imagine having your life be ten times better? I can imagine my friends and people in need having ten times more and having a much better life. This is one of my favorite reasons for bringing virtual volunteering into the nonprofit scene. Just like Henry Ford was able to pull people off the sideline and help them make ten times more, there is so much wasted time among passionate people with the skills to be incredible e-volunteers.

Sliced Bread and Volunteering

The Earl of Sandwich was an early pioneer in efficiency when he asked for a meal he could hold. However, did you know it took centuries before sliced bread was created? Sliced bread finally made its appearance at the turn of the 20th Century! But no one bought it. Sliced bread appeared. Sliced bread was ignored. Sliced bread disappeared. Sliced bread is sort of like virtual volunteering. It just takes a while and the right timing.

Isn't it incredible that there are so many things that are "the greatest invention since sliced bread," but sliced bread wasn't such a great invention since people ignored it for 15 years! For 15 years, no one wanted sliced

E-volunteer Revolution

bread! Well the Web 2.0 has been around for over 15 years, and it's about time for virtual volunteers to power nonprofits to new frontiers.

We're changing how people view volunteering. It's not like "Everything you think you know about volunteering is wrong," but it is closer to "Everything you think you know about how to make a big impact on the world is wrong." But honestly, you probably know what it takes to make a big impact, and you know the dual importance of using the Internet and having a passion for something. All it takes is the logical next step of virtual volunteering for nonprofits.

Nonprofits and volunteers need to jump into the virtual game. It's probably best to do it together. Yeah, it's definitely best that nonprofits and volunteers (and while we're at it, let's add passionate, socially-minded startups) all jump into virtual volunteering together.

So here's the key for collaboration and virtual volunteering: the virtual skill needed and the mission of the nonprofit/project have to match with the virtual volunteer. Apart from the Internet this is pretty hard, but this is the only way.

Please don't fall into the fallacy that someone who works from home on a laptop can do any specific virtual skill, programming language, or operate a software purchase you made! That sounds crazy, but that's what people ask for all the time. Your tech friend is excellent at only a few things, and while he or she could spend hours solving a problem that would take you ten hours to solve, it's probably better to pay someone who can solve your issue in 15 minutes if you can't find a free e-volunteer outside your personal network. Reserve the people in your personal network for the things they're best at. (Tip: you can also list things you need help with and make your personal network aware of them while the deadline is still over a week away. This gives them time to consider whether they want to learn about something new which will take a significant amount of time. So please don't force your virtual volunteers into a time crunch if you can avoid it.)

Remember that matching the skill and the passion is crucial for growing virtual volunteering. You will get a lot of good tips throughout these pages for narrowing your passion and isolating the specific virtual volunteer skills you need to make a big positive impact on the world.

Small Fish in a Big Ocean

The digitalization of skills, tasks, projects, jobs, careers, and companies means that globalization now allows access to talent and virtual volunteers around the world. The things that nonprofits and volunteers want to do to make the world better—all those steps, projects, dreams—can be assisted by virtual volunteers from anywhere in America or anywhere in the world. In globalization there are winners and losers, but the goal of digital altruism is for everyone to benefit. There are enough people with e-volunteer skills to make sure no one's quality of life decreases.

Due to globalization and digitalization of the workforce and consumers, the talent pool is a lot bigger than it used to be. Also, everyone has some online skills to donate as a virtual volunteer. Your grandmother on social media can virtually volunteer to share what she considers to be a nonprofit's good posts and videos. This small pebble of virtual volunteer support can help reach over the course of a year between an additional 1,000 to 30,000 people for the nonprofit's awareness and cause. This has real-world impact. This tiny activity of a person with the most basic virtual volunteer skills sharing a nonprofit's posts once a month can make an outsized impact on the world—most likely by making someone in her network aware of a specific cause and project that is a perfect fit.

That is just the tip of the iceberg, and all it requires is someone saying they have enough passion for a nonprofit to share it with their friends once a month. There is a huge spectrum of virtual talents and abilities. People might live in the poorest countries, but if they have a laptop and internet connection they can and are starting companies. These same gifted and hardworking women and men are supporting nonprofits with their computer programming knowledge and expertise. The world is busy building tech companies and virtually volunteering. Americans, especially, need to both virtually volunteer and seek out local and global virtual volunteers. Globalization, especially in its digital phase, is favoring the doers, the risk-takers, and the collaborators. It's better to be a small fish nonprofit jumping into the big ocean than to continue to be a small fish in a tiny pond or fish tank. You can only grow so big in a small fish tank.

Finally, It's Easy to Do Good

Another big winner in the digitalization of altruism is each person who wants to make the world better. Virtual volunteering is making it easier to do good for others. Technology, algorithms, and machine learning are helping humans enjoy volunteering more, make a bigger impact, and volunteer more for the things they care most about. The new ease of virtual volunteering is probably one of the most crucial aspects which goes unnoticed. Humans prefer the path of least resistance. This is just part of being human—conserving energy. When making a positive impact is easier, more people help. When making a positive impact risks not succeeding or being mocked by peers, then people overwhelmingly choose to keep the status quo. Therefore, with the phenomenon which is virtual volunteering, one encounters a sea change in how altruism is perceived and achieved. Of course this is natural with the advent of computers and social networks, but it is also unnatural to human altruism because it is digital. Walking this fine line is key to making altruism easy, fun, and rewarding. The gap between the unnatural digital versus the natural altruism must be crossed to usher in digital altruism. The nonprofits and socially-minded businesses and individuals who can perfect this will be the biggest winners in growing virtual volunteering for their causes and the world's betterment.

Businesses' Vulnerability Is a Strength

On the previous note of socially-minded businesses and individuals taking advantage of the virtual volunteer revolution, for-profit entities have perhaps the most to gain. It is ironic that businesses are forced into the dual roles of villains in the present and Cinderella stories in the past.

Let me explain. Businesses when they start out—whether a local restaurant or a tech startup—are highly

E-volunteer Revolution

reliant on volunteer help—physical, financial, and virtual. These new companies are just a dream in someone or a couple people's minds. They ask for help; they make some sales. They get help first constructing and repairing a food truck or they get much-needed introductions to crucial vendors or clients. The new business owners/dreamers ask for help—beg for help—because they know they need it. Some of the biggest billion dollar companies had to ask people to buy cereal boxes from them to keep their dream alive, and people supported them because they supported the dream/mission. Their supporters saw the hard work and dedication of the founders.

As businesses become more successful they stop asking for help. They don't need to ask for help, and this is a good thing. They shouldn't forget the help they received, and the public shouldn't forget the blood, sweat, and tears—mostly of the founders—which went into the company. The good news is that most companies that make it to profitability are companies which do more positive than negative for the economy, consumers, and tax payers. Everyone benefits.

Those companies asked for help, and now that they are stable, successful, and growing, many are helping new companies. The majority of e-volunteering is being done business-to-business, and nonprofits are not involved. (Hint: many nonprofits are not involved in getting e-volunteers due to ignorance or time-constraints. That ends with the e-volunteer revolution. That ends because all the people who made the biggest difference in our lives did it "off the clock.") Companies are giving back both financially and with e-volunteer skills. Successful companies are run by founders, upper management, middle management, and skilled employees who are volunteering to help others in their industry and those who align with their passions. There is a lot of free networking, consulting, volunteering, and virtual volunteering going on behind the scenes.

The winners are the dreamers who are making detailed goals to make sure their dreams are structured goals not wishes meant more for genies and lottery tickets than for virtual volunteers. The winners in the new age of virtual volunteering know exactly what services they need and are asking for specific help. The winners are letting virtual volunteers know how the project, nonprofit, or new business is making the world better. No one wants to help you swindle someone else, but if what you're doing makes the world a lot better—whether it's finally getting good Mexican food in Kentucky or getting everyone more access to online education—if you can make the world better, people will help.

Good Communicators

The last group of winners I'll highlight here are the good communicators. Work on your communication for your passion. Communicate what you've done to this point. Someone who has passion but no progress does not have enough passion. In fact, their passion lies elsewhere—which is okay if they make that passion known. Someone might have passion for fast cars but also have the tech skills to make adoption safer for the kids, less expensive for parents, and better for everyone who actually cares about the well-being of the children. It is natural for humans to develop passion in accordance with their skills and their needs. America's most famous

E-volunteer Revolution

visionary for smartphones, tablets, laptops, and fonts had a passion for yoga. But that passion for yoga faded in priority yet also increased in impact as he built the world's most valuable company.

So be honest about your passion. Be honest with yourself and others. Make sure you are not lying to yourself about having a passion for fast cars when what you really want is others to respect you and for others to be able to spend a semester studying abroad in college. Fortunately, most people reading this know their top one or two passions and can differentiate that from their hobbies. But I'll add, don't change what you're passionate about depending on whom you're talking with. It's human nature to alter what you claim is your passion in the presence of someone deemed a potential benefactor or partner. Don't change who you are! Don't change your passions from conversation to conversation. You have a main passion; don't falter. You can change how you describe your passion, but be yourself and invite people to join you in making the world a better place according to the mission you've set out on. It's hard feeling alone in your passion, but changing for others won't help you accomplish what really matters to you—you'll just waste your time and others' time.

On that note, communicating your passion is just the start. To get help from volunteers and virtual volunteers, you need to be able to communicate the specifics you need, when you need them. You need to also communicate the specific benefits the volunteer will receive—usually this is a report of the success and impact of the volunteer's effort. The winners in the virtual volunteering revolution will be able to communicate well. However, this is not so simple.

Big Shifts in Volunteering

No one could step onto a sailing ship for the first time and act as captain or first mate to direct the crew what to do, when to do it, and to what extent. Asking for help from virtual volunteers is the same. No one can communicate what they need or whom they need help from if they do not know the lay of the land. Good communicators are like good contractors who know what needs to be done and how to do it. The winners in organizing virtual volunteers will know the terrain well enough to make specific requests. Those who can't make specific requests will ultimately waste virtual volunteers' time and be unable to reap the benefits of the virtual tech cavalry standing by to help as needed.

Chapter Three
Surprising Stats

Nonprofits should be requesting 90% e-volunteers and only requesting 10% physical volunteers not vice-versa.

Businesses are going virtual at an alarming rate. Nonprofits are sitting on the sideline at an alarming amount!

I could write a four to six sentence introductory paragraph, but it seems best to let the statistics speak most in this chapter.

The Majority of People Want to Work Remotely

2016 was the first year that the majority of U.S. workers preferred to work remotely, off-site, or from home. The majority of people want to work remotely and virtually. This means a few things for volunteering.

Surprising Stats

First, volunteers are people. People want to work remotely. Volunteers enjoy having the option and freedom to work remotely and virtually. Give it to them by finding ways they can virtually volunteer for your cause and mission. Second, there is a virtual volunteer army already out there in massive numbers. The majority want to work remotely even though they are not all being given that freedom from their employers. Virtual volunteering is a great way to give them some of the freedom and autonomy they are seeking. Third, the American workforce is developing more and more remote and virtual skills. This is helpful for both their ability to complete tasks in a virtual environment and also to be good communicators when face-to-face communication is not an option or not realistic. America's virtual volunteer army is improving their skills and waiting to be called upon. You don't have to convince anyone to e-volunteer. They all already want to e-volunteer!

Nonprofit Discrepancy

28% of businesses depend on remote and virtual work. Less than 1% of nonprofits rely on virtual volunteers. Businesses are continually seeking to increase in efficiency, employee satisfaction, customer satisfaction, and product development. Businesses are going virtual at an alarming rate. Nonprofits are sitting on the sideline at an alarming amount! I hate to say this, but nonprofits used to be led by visionaries, and it is time for nonprofits to remember that they are crucial to making the world better. Making the world better requires more than physical effort, it requires wisdom to pursue technology that can augment the physical effort. Nonprofits need to close the gap. 1% compared to 28% is not acceptable.

As I've researched and interviewed nonprofits I've found the 1% who do allow virtual volunteering are not even seeking to grow beyond 1%. These nonprofit organizations are only allowing e-volunteering because some volunteers have allergies or phobias that keep them from volunteering in the physical office. Therefore, the real number of nonprofits taking advantage of virtual volunteers is closer to 0.

Nonprofits Are Ignorant of E-volunteering

Less than 10% of the current volunteer opportunities being requested—that is requested not needed—can be done remotely or virtually. Nonprofits are overwhelmingly seeking in-person volunteers and not considering the potential of requesting virtual volunteers. Nonprofits should be requesting 90% e-volunteers and only requesting 10% physical volunteers not vice-versa. Nonprofits need more knowledge about what is possible when they take their mission digital.

Minimum Wage E-volunteer Tasks Make up 99% of Requests

While only 10% of nonprofit requests for help can be done virtually, less than 1% of virtual volunteer requests would pay more than minimum wage.

Surprising Stats

Nonprofit organizations are not asking for high-level or technical virtual volunteer help. Unfortunately, nonprofits are asking for the same things that the onsite volunteers are already doing. For example, nonprofits are asking for someone to type out their handwritten notes from a meeting instead of asking for setting up a new customer relations management system, creating an online campaign, organizing a fundraising event, etc.

It should be pretty simple. There are 10,000,000 highly skilled tech workers in America making $40, $70, $100, or more per hour doing specialized and varied skills. Learn how they can help you and ask for specific help. There are an additional 30,000,000 American workers with sufficient tech skills who can also help. This does not even take into consideration the students. Find a way to make virtual volunteer tasks available that would normally cost $50 to $115 per hour, and then allow e-volunteers to fill those rolls.

Physical Volunteering Is Trapped Under Minimum Wage

The experts on volunteering and efficiency have been clear that the majority of volunteering in-person to serve food, arrange food, deliver goods, paint buildings, etc. is trapped below minimum wage. This means that the value being provided by the volunteer is less than the volunteer could live off of it that was necessary.

Of course there is significant social value from volunteering, and many people volunteer in order to do something fun, to give back, and to do something different than their day job. However, nonprofits can do a better job at helping make volunteers' efforts have a higher return on time invested. Volunteers can also take the job and tasks more seriously when appropriate. Volunteer opportunities are supposed to be fun not business-like and stressful. However, businesses are making their work environments much more fun, virtual, and rewarding. Nonprofits can do the same!

Companies and Nonprofits That Get Help Make the Biggest Difference

Less than 0.19% (that's about 1 out of 1,000) new ventures receive orchestrated official help commonly referred to as venture financial, knowledge, and network capital. However, although only accounting for one out of a thousand companies, they contribute to over ten percent of jobs and over 20% of GDP. That means that these companies that get orchestrated help and support make a 1,900% increase in the nation's economy. That's the power of organization, passion, collaboration, and support. Businesses usually have to pay for these, but an organized nonprofit can ask for specific help from virtual volunteers and should ask for help.

New business startups are able to make a 1,900% increase so quickly only because of the combined powers of globalization and digitalization across industries. The same benefits that have been powering the private sector can be applied to the nonprofit, charitable, educational, religious, as well as socially-focused small businesses and startups. Orchestrated help is possible for the local nonprofit not just the hottest tech startups. Orchestrated digital help is called e-volunteering.

The Workforce Is Ready to E-volunteer

Over 60% of the U.S. economy is currently in the service sector and intangibles. Less than 30% of the U.S. economy is now in measurable production units as found in manufacturing and agriculture. The 60% of American workers in the services industry are using more and more technology that can be directly applied to nonprofits. Every person can volunteer, but for the first time the majority of potential volunteers are developing skills to contribute specialized virtual volunteer hours. Now every volunteer can be a virtual volunteer—able to help whenever and wherever with causes they're passionate about.

The Creators Are Few

About 10% of humans create 90% of the content, jobs, and well-being in the world. Regardless if this is true, it is true that all humans share a healthy level of altruism and curiosity. All humans are on a creativity spectrum. Some might create more, and some might consume more. Nevertheless, the freedom to create is essential, especially when some creators will withhold their creativity in the face of obstacles. We need to help remove obstacles so that everyone is free to contribute, and virtual volunteering provides a way for people to create and make an impact that otherwise would never exist. What I'm saying is that the e-volunteer revolution needs humanity's creators to have patience along with passion. True creators want to empower others to be creators. I believe that with all my heart.

Chapter Four
Big Secrets in Volunteering

It does not make sense for a business to feel freedom to ask for volunteers to help sweep two feet of mud out of the building but not feel free to ask virtual volunteers to update the company's online presence.

Jobs Matter

The turn of the millennium saw a huge shift in global priorities. In the past, the majority of people prioritized religion or freedom. Now the priority is a job. It's not having a family, democracy, a home, or even heaven. The priority of the world is having a good job. This might be odd for Americans who have never had to worry about having access to a good job, but many in America and most people in the world prioritize having a job and income.

E-volunteer Revolution

This is a shift, and it's a good shift in a lot of ways. Incomes provide freedom, worth, opportunities, and self-determination. Most importantly, providing someone with a job is better than providing them with charity. Teaching a person to fish is better than giving a person a fish. The difference between now and the past millennia is that everyone knows it!

Everyone knows that giving a person a skill and a job is better than giving a person some money. Everyone in the world knows that having a job, having job security, and being able to get another job if necessary is crucial. Everyone seems to know this except volunteers. This is unacceptable. It is our responsibility to let others know how to do more good. Volunteers want to make the world better and go through great lengths to donate their time and money.

If you want to help someone in need, then help that person with what they want most: a good job. If you want to help someone in need, then help that person with what the experts say that person needs most: a good job! If you want to help someone in need, virtually volunteer for a nonprofit or a business that will play a crucial role in getting people jobs and getting people plugged into the economy. This is so important that there isn't anything else to say. That's the point of altruism, volunteering, and e-volunteering.

Okay, I'll say it again—really. If you want to help someone in need, then help that person with what they want most: a good job. If you want to help someone in need, then help that person with what the experts say that person needs most: a good job! If you want to help someone in need, virtually volunteer for a nonprofit or a business that will play a crucial role in getting people jobs and getting people plugged into the economy.

Big Secrets in Volunteering

Each $1 of Profit at a Good Business Becomes $50 for the Economy

I'm going to trust you with one of America's biggest secrets. Leaders know this secret but don't trust people to understand it. The secret is that for every dollar of profit that an honest business makes, fifty dollars are contributed to the economy. A dollar of honest profit is multiplied by fifty in the economy! This is why businesses, enterprise, trade, innovation, etc. are so crucial to a strong and prosperous economy.

The reason for this is because profit is not easy to make. Sales are one thing, but clearing a profit takes ingenuity, knowledge, risk, and hard work. Profit at a company that plays by the rules is not easy. Profit comes from adding value to the economy, and profit leads to more hiring, salaries, and products at a good company. Profit at a company comes from making people's lives better—that's why consumers give their money for a product or service. In-demand products and services create ripple effects that bless everyone in the economy. Consider for instance when American cars began to get a bad reputation for breaking down while foreign cars were lasting 100,000 miles longer without expensive repairs. Over several decades, the loss of value and loss of profit at American car manufacturers had negative consequences for everyone connected to the car industry and everyone in the economy—since people in the American car industry spend their money in the American economy. Inversely, foreign economies producing higher quality cars with a higher perceived dollar-for-dollar value saw their entire economies benefit from the principle of $1 of profit translates into $50 for the local economy.

My concern is less why leaders don't trust people to understand the fifty times multiplier of profit. My concern is that virtual volunteers and nonprofits understand it. The quality of life and the freedom that nonprofits and volunteers want for people requires nonprofit organizations and businesses to work together. That is why virtually volunteering for businesses is crucial for the global economy. The world is not a zero sum game. Technology, peaceful dialogue, efficiency, and collaboration can increase the quality of life for all people with no one left out. Charities are seeking, finding, and helping those who have been left out. It's up to businesses to make a place for these people to return to which offers good employment, freedom, and advancement. I don't believe this incredible opportunity was possible prior to the digital age, and that is why I am so passionate about virtual volunteering.

Volunteers' Fading Value and Growing Fatigue

By 2007 the term "Katrina fatigue" entered the American consciousness, and over 50% of Americans had Katrina fatigue. Katrina fatigue is mainly described as the American populace growing tired of hearing for over 2 years about how bad the situation continued to be in New Orleans. Sometimes Katrina fatigue is also associated with Americans trying to solve issues like racism or systemic poverty though financial aid as a hopeful, easy solution.

Did you experience Katrina fatigue? American volunteers responded powerfully to the dire situation left by governmental breakdowns at every level in the

Big Secrets in Volunteering

Katrina aftermath. Volunteers drove across the country to help with the physical cleanup, repairs, and rebuilding of the Gulf Coast. Volunteers spent an average of 3 days per volunteer sleeping in churches at night and doing whatever was asked of them during the day. Volunteers really gave a lot of their time to help. However, fatigue can set in among responders as well.

It's sad in the aftermath to see volunteers take any of the blame, but they did receive their fair share of criticisms. Two years of nonstop Katrina fatigue-causing news stations constantly required new stories and new analysis. Some of the ire of the analysts fell on how much the volunteers were spending to drive to Louisiana, miss time at work, and not accomplish enough "value." It was actually said that volunteers cleaning out and repairing homes were only worth about $5 per hour. It was suggested that the volunteers should have stayed home, earned their regular income, and sent much more money to Katrina victims. This just isn't reality.

The volunteers were taking their PTO and vacation days. Going to get a second job just wasn't a realistic option. People volunteer because they enjoy it not because they're forced into it. Furthermore, virtual volunteering wasn't much of an option at the time since Katrina came ashore before smartphones became popular in America. Of course, all this goes to show that nothing exists in a vacuum. Volunteering is the altruism in human DNA. However, virtual volunteering is just another form of volunteering and efficiency that's currently available due to humanity's technological progress.

The most recent hurricane disaster caused by Harvey is estimated to have cost the Gulf Coast 180 billion dollars and ranks as the costliest natural disaster on record. The economic recovery is expected to take a decade. Houston and the Gulf Coast's nonprofits and businesses need to call on virtual volunteers and ask for help. It does not make sense for a business to feel freedom to ask for volunteers to help sweep two feet of mud out of the building but not feel free to ask virtual volunteers to update the company's online presence.

Corporations Use Virtual Volunteers

The biggest corporations in marketing, big data, entertainment, and social media are using virtual volunteers. Usually the virtual volunteer doesn't know it or doesn't mind. When people fill out surveys, submit reviews, allow location tracking on their phones, allow auto-reporting of software glitches, and do many other seemingly harmless things, big corporations are benefitting—sometimes to the end-users' long term detriment. It's easy for big corporations to get users to virtually volunteer a few minutes of time or to volunteer their data because corporations make it easy and seamless. Corporations will continue to do what they deem necessary to gain a competitive advantage. The best response is for people to start virtually volunteering for causes they care about by helping nonprofits and local businesses which give them that outlet.

Fast, Cheap, and Good

There are three important things that are impossible to have at one time: quick, inexpensive, and quality. Businesses know that they can only have two of the three. Fast and cheap but not good. Cheap and good but not fast. Good and fast but not cheap. Any business or organization that can do things quickly, cheaply, and with high quality will succeed. Until the digital age, it was impossible. Even with the digital age, it is impossible for most businesses.

Fast, cheap, and good is only possible with the combination of altruism and the digital age. Virtual volunteering makes it possible. The secret is that virtual volunteering tricks the "cheap" part of the equation. You can get $100 and even $500 per hour of technical work, consulting, or connections for free because the virtual volunteer cares about your cause and can meet your specific need. When you use technology to organize your needs and to sort your virtual volunteer support then you can accomplish things quickly, cheaply, and with high quality. You can jumpstart your mission to make the world better with virtual volunteers.

Panic and Peace

Companies with big profits don't panic. Nonprofits with large endowments don't panic. When there is a war chest or a buffer, there is peace instead of panic. When there is peace there is freedom and creativity. Businesses and nonprofit organizations under extreme stress cannot change the world. Nonprofits with a lack of funding can feel an existential threat that makes them focus on fundraising instead of focusing on making the world better. What if you didn't need funding for all the tech because the tech was all free? Can you imagine the possibilities? Virtual volunteering can help with this by bringing in extra resources to restore some margin, rekindle some dreams, and give some peace of mind that the bottom is not going to fall out. There is a wealth of skilled e-volunteers ready to help good causes succeed and restore peace and creativity to the visionaries.

Freedom and Results

The results-only work environment (ROWE) movement is growing in popularity because it provides freedom. If an employee can deliver the requested results, then the employee is free to skip meetings, work from home, and do whatever he or she desires to achieve the results. This is crucial for volunteering because volunteers need to be given freedom and be given specific tasks that can lead to results. Virtual volunteers can get the work done, but they're adults not kids. Give them the requirements and the freedom to get it done.

Businesses have seen tremendous success by focusing on providing freedom to achieve results instead of pressure or financial reward to incentivize results. Employees who spend over half the workweek away from the office were the happiest. Employees who spend over half the workweek away from the office have better relationships at work and are more likely to name a co-worker among their inner circle of best friends. People want autonomy and purpose. Virtual volunteering is a perfect match. (Tip: freedom not money has been found over and over again to be the best incentive. You can implement more freedom in your nonprofit for volunteers to try their own ideas when you have more support from virtual volunteers. More freedom leads to more volunteers not less volunteers. It's a win-win.)

Big Secrets in Volunteering

The Wealthy Volunteer

The wealthiest and most successful people volunteer five to ten hours per month. The most successful people only volunteer at one nonprofit or institution, and they sit on the boards and are highly invested in the success of the nonprofit organization. (Hint: the wealthiest volunteers maximized their focus prior to the digital revolution by choosing one nonprofit; think of how to maximize people's efforts with the potential of the digital age!) The wealthiest and most successful benefactors and volunteers have laser focus on how they can help a cause they care deeply about.

The wealthiest people also know that volunteering is good for their physical health. They don't have time to volunteer but they make time to volunteer. The wealthy also know that running or exercising daily will add hours to their day and add years to their life—not the opposite. Successful people have always been able do a lot of things that others wish they had the time to do. The unique thing about successful people is that they weren't always successful and they didn't always have the time to do what they wanted to do. They found ways to make the time, and you can too!

Virtual volunteering allows everyone to volunteer. Virtual volunteering doesn't require being at board meetings at a certain time, driving anywhere, or making a set schedule. Virtual volunteering allows anyone to receive a digital task or project and complete it. Virtual volunteering allows teams of friends to complete a project together. Virtual volunteering allows an individual to improve her physical and mental health by contributing to the well-being of others. Altruism will make you feel good—it's in your genes. Also, a virtual

E-volunteer Revolution

volunteer with an empty bank account can still donate virtual skills and have a bigger long term impact on the world than a wealthy businessman with the title trustee. Virtual volunteering has limitless potential.

Chapter Five
Bad News: Inefficiency

The same way the Internet created a new paradigm for businesses, virtually volunteering is harnessing the Internet for altruism and humanity.

When did the private sector start doing over a thousand different practices that have not been integrated into the nonprofit sector?

Nonprofits and small businesses are hoarding and relying solely on their own small group of talent and supporters. This is growth-stunting for the nonprofit organization and disastrous for the local ecosystem.

This is not a chapter I want to write because I hate that nonprofits are underfunded and understaffed. I hope more nonprofits will see that virtual volunteering can help increase the amount of tasks accomplished and make sure funds go where absolutely necessary. So there is some bad news in this chapter for nonprofits, visionaries, startups, and the economy, but I'll make it quick and the good news is just around the corner in the next chapter.

Untapped Skills and a New Paradigm

Too much skill is being left unused. People have skills, knowledge, and abilities that can be tapped into by nonprofits and businesses via virtual volunteering. Everyone is at fault and no one is at fault. People will usually do what they consider to be in their own best interest. Therefore, when people aren't at work, they do the things they enjoy doing or must do. Many times people will spend their leisure time semi-relaxing watching TV or on social media. It's well known and even publicly admitted by TV studios, movie production companies, and social media companies that this semi-vegetative consumption is not good for mental health or physical health. We humans do it because it's natural to rest with our curiosity on autopilot (Hint: it's called channel surfing or scrolling through social media.) In America, there are 40,000,000 well-qualified virtual volunteers with an unsatisfied feeling because their ability to create and contribute is not being empowered. There is double that number of people with skills to contribute and grow via e-volunteering. Every year, trillions of dollars of economic improvement go untapped because the current system makes it easier to drone on autopilot than to make an impact on the world.

What we all need is better rest such as quality sleep, and we need to take the reins of our own curiosity. Virtual volunteering allows for that, but everyone is going to have to work together to make the process of virtual volunteering easier. Volunteering is like running and exercising—it's one of the best things for you but very few people can stick with it on their own without support. We need to continue to develop e-volunteering's best practices and build up the sector of virtual volunteering. The untapped potential is a problem, but it's a problem that can be solved by creating a new paradigm. The same way the Internet created a new paradigm for businesses, virtual volunteering is harnessing the Internet for altruism and humanity.

Untapped Online Terrain

The economy is such a physical thing in our minds but it doesn't have to be. In the 1980s and 1990s the market economy in the United States was described as magical by the leaders of the Soviet Union. The communist leaders knew communism was doomed when they compared the biggest Soviet grocery store with 1,000 items to the biggest American grocery store with almost 60,000 items for sale. The communist leaders didn't understand the magic, but they knew it was real. The magic is that allowing people freedom to buy and sell what they want at a price they agree on works. 300 million people can organize an economy on their own without management from the top.

If you look around America, you will see that everything physical has a price and has an owner. Let's put political disagreements aside and look at the facts. Personal property rights in America means that everything you see has an owner. 300 million owners of billions of physical items equates to everything being used at the maximum efficiency possible under the market system. There is no frontier land in America you can go claim for free. Every square foot of office space and residential space in New York City is claimed by someone. There is nothing left that is physical.

Enter virtual volunteering and the digital age. The Internet allows for new frontiers and new creations. The costs for digital space and data transfer are so small that they are almost nonexistent. Hosting and servers cost next to nothing when a nonprofit or business is providing value and bringing in money. The digitalization of globalization makes the online world an ever expanding economy. It's time for nonprofits to take advantage of that. The costs to build new things, to test new projects, and to make an impact is easier and faster than ever before! It's time for virtual volunteers to see that they are on the precipice of an adventure. The opportunity is right there. It's the e-volunteer revolution.

Nonprofits Can Waste People's Time

There's no other way to say this: nonprofits waste volunteers' time. Fortunately, volunteers are aware that inefficiencies in nonprofits' offices are the "nature of the beast." However, virtual volunteers believe that the digital age changed the nature of the beast. It's time for nonprofit organizations to catch up.

Many times nonprofits reserve menial tasks for onsite volunteers. Volunteers who make $30 to $50 to $100 per hour at their jobs often feel like nonprofit employees mistreat the volunteers as children instead of as successful adults when they volunteer onsite to offer physical labor. Honestly, this is probably as much or more the fault of volunteers who are not taking their service seriously enough, but it also has a lot to do with the aforementioned "nature of the beast" and fatigue from the unending war on poverty. Overall, I believe the problems stem from nonprofit organizations not being free to focus on their core competency and core mission. I would not have much hope for improvement if not for the potential for e-volunteering to free up nonprofits to work more efficiently and accomplish more in the future.

Too Tight a Grip on Suboptimal Resources

Nonprofits and small businesses are hoarding and relying solely on their own small group of supporters. This is growth-stunting for the nonprofit organization and disastrous for the local nonprofit ecosystem. Due to being underfunded and understaffed, nonprofits, individuals, and even small businesses seek money/sales/donations/investment/service/patronage and support from their small circle or network. This is a problem because it stifles the matching of passion and skill. When you trap talent in your circle you hurt others.

All people have skills and passions. People have to be able to volunteer for a cause they are extremely passionate about and feel they can make a difference in. When a nonprofit hoards volunteers who are only semi-passionate about the nonprofit's mission, then the other nonprofits who champion those volunteers' passions miss out on much needed and effective help. The volunteers miss out too. This is the system that makes volunteering ineffective. This is the system that shouldn't exist in the digital age. Virtual volunteering can help all these nonprofits be more effective, better funded, and have much happier volunteers.

Ignoring Advances in Businesses

When did the private sector start doing over a thousand different practices that have not been integrated into the nonprofit sector? Businesses are using an array of software, strategies, data, and technologies that the majority of nonprofits don't know about. How is this even possible? How is it possible when there are free versions of all these technologies which are crucial for making the impact which nonprofit directors and donors dream of making?

It's possible and happening because the digital age creates and perfects digital tools so quickly that it requires companies to make a lot of profit just to be able to afford to stay caught up and in the game. 99% of nonprofits don't have that kind of revenue stream. However, what nonprofits do have is access to all—every single one—of the employees who work at these billion dollar businesses, Fortune 500 companies, and impressive local businesses. Best of all, these companies want their employees to virtually volunteer.

Chapter Six
Good News: Potential

The U.S. economy has the world's top ecosystem of talent/labor capital. Businesses are taking advantage of it. Nonprofits and socially-minded businesses need to ask for help. Everyone is right there waiting for someone to give them a good reason and give them a promise to not waste their time e-volunteering.

Let's get right to the good news and look at the potential of collaboration made possible by technology.

Smallpox Eradication: A Pathway to Follow

The horrific tragedies, vaccination, and eradication of smallpox present a pathway for virtual volunteering to follow in fighting poverty and disaster. Just like with smallpox, the answer is in technological advancement, leadership's public acknowledgement, and collaboration. The results will be at a magnitude beyond imagination. This is the power of altruism in the digital age.

Smallpox is the worst virus humanity has ever encountered. The smallpox virus became deadly to humans centuries before the birth of Jesus of Nazareth. While the Black Death killed 75,000,000 people in Europe, in the 1900s smallpox killed 300,000,000—over four times the amount of the Black Death. Smallpox kills a third of the people infected, 80% of children, and leaves a third of the survivors blind. Smallpox killed 90% of the Native Americans when Europeans brought the disease to the Americas. Smallpox has killed over a billion people. However, thanks to technological advancement, leaders' public acknowledgement, and collaboration, the smallpox vaccine has now saved over one billion lives.

Edward Jenner created the smallpox vaccine in 1798. Jenner found that inoculation with cowpox provided protection from smallpox—the Latin word for cow is "vaccine." However, having a vaccine is different from getting the vaccine to everyone who needs it. This is the same case with virtual volunteering bringing altruism into the digital age. Hundreds of millions of people continued to die of smallpox after the discovery of the vaccine. Even during the century that split the atom, went to the moon, fought two world wars, and saw the rise of communism—over 300 million people died of smallpox in the 1900s despite there being a vaccine.

The first well-known person to risk pursuing global altruism was Viktor Mikhailovich Zhdanov Deputy Minister of Health for the Soviet Union in 1958. Perhaps someone else would have challenged the world to eradicate smallpox in the poor countries being devastated by the disease. Perhaps not. He took a risk and challenged the world.

Out of all the biological plagues on humanity, smallpox alone has been eradicated. But the same type of collaboration can lead to the 100% eradicate polio, measles, malaria, and Guinea worm—it's just not easy with billions of people. Zhdanov followed Jenner in doing his part to be a hero—unknown to almost all he helped save.

The final stage in the eradication of smallpox and the preservation of over a billion lives to date is Donald "D. A." Henderson who was the head of the World Health Organization's successful eradication program from 1966 to the last case of smallpox in 1977. Henderson is well known for his altruism and also for his passion to see altruism done effectively, efficiently, and with collaboration. Henderson had all the traits necessary to save a billion people. He was a risk-taker, visionary, ambitious, diplomatic, and blunt when needed. He could never have done it alone, and the people in poor villages responded in ways that shocked the world's leaders who did not think Zhdanov and the communists' goal to eradicate smallpox was possible. The people did it. The poorest villagers carried victims and information through thick jungles and dangerous areas. With organization, tasks, freedom to operate, passion, and technology, the people did it.

Virtual volunteering is the same. We still have the main diseases like malaria, but even more so, the disease of poverty persists. Academics don't know what poverty is, but people in poverty and people under oppression know what poverty is. It's true that the poverty today is not as bad as poverty in the 300s, 800s, or 1600s. But it's also true that poverty, oppression, and the inability to control their own lives affects billions of people. No one likes this—especially the people in poverty and the empathetic people who seek solidarity with them. However, despite humans' altruism, humans also have a significant survival instinct that can overrule altruism with apathy. Humans are altruistic, and that altruism needs to be nurtured with virtual volunteering.

Nonprofits, individuals, governments, churches, and businesses are doing what they can to make the world better. Too many are doing it without the power of the digital age and virtual volunteers. It doesn't make sense. Sort of like 300,000,000 people dying of smallpox in the 20th Century didn't make sense. Volunteering hasn't yet been brought into the digital age. This is probably because world leaders didn't have a childhood with personal computers, social media, and efficiencies provided by online technologies. Virtual volunteering has its digital-age creators just as smallpox had Edward Jenner. What virtual volunteering lacks is a clarion call to action just as the unlikely communist leader Viktor Mikhailovich Zhdanov sounded the death knell of smallpox from behind the Iron Curtain. Virtual volunteering also needs a D. A. Henderson to organize and trail blaze the movement. Fortunately, though, this is the digital age and there will be and already are millions like Zhdanov and Henderson already swelling the ranks and effectiveness of virtual volunteering.

Human Capital

It is common knowledge in the business world that the United States' human capital is over ten times the amount of the United States' total physical capital. When a big company acquires smaller companies it looks

Good News: Potential

for two main things: the physical capital of the company and the knowledge capital of the company. Some companies are acquired not because they have so many customers, revenue, new technology, or patents. These companies are acquired because they have really good employees, excellent managers, a good company culture, and what is referred to as incredible talent.

The U.S. economy has ten times more talent/labor capital than physical capital! That means that people have knowledge, skills, and experience that make them very valuable. When a person, a group of people, and even an entire labor force have labor capital, there is tremendous potential. The U.S. economy has the world's top ecosystem of talent/labor capital. Businesses are taking advantage of it. Nonprofits and socially-minded businesses need to ask for help. So many people are waiting for someone to give them a good reason and give them a promise to not waste their time when they choose to e-volunteer.

Don't Waste Money

Nonprofits have tight budgets that they have to stretch. There is no better way to stretch a budget than to replace parts of the budget with free volunteer labor. Nonprofits are not or should not be paying someone to vacuum the floors—maybe or maybe not clean the toilets—but not to vacuum the floors. (Tip: a nonprofit director or business owner must have the humility to clean the toilets.) Donors expect their money to be used wisely, and nonprofits do use it wisely and stretch that money so unbelievably far. So it is time for online tasks to be outsourced to virtual volunteers. The main obstacle will be the nonprofit organization's ability to identify and manage the tasks and projects. More on that in the chapter for nonprofits.

Creatio ex Nihilo

So perhaps e-volunteering isn't creating out of nothing, but in some ways it is. When you're sitting on the subway, on a bus, or on your couch there are a lot of ways to virtually volunteer with just your phone—no calls necessary. We have all seen how in public everyone has their phone in front of them—while at dinner, while on the subway, while walking down the sidewalk, while crossing the street, and while driving. Don't text and drive. (Tip: what I do if I need to send a text while driving is pull over into a parking lot—even if I have to exit the freeway. We all know it isn't worth it to text while driving, and we need to think about the other drivers' lives we're risking not just our own lives.)

So people text and drive. They do this because they want to get more done. People have their phones in front of their eyes all the time because it lets them "get more done" or stay connected better. Their phones allow them to do what the human mind wants to do for survival and social status: consume and connect.

Virtual volunteering allows the altruistic side of humans to create, consume, and connect in a way that helps others. Virtual volunteering can be done anywhere and on any type of device. Virtual volunteering goes hand in hand with physical volunteering, altruism, and being human in the digital age. To this point, virtual volunteering is not a part of people's daily lives. To make the world better, e-volunteering needs to be easier for all.

Good News: Potential

The Serenity Prayer in the Digital Age

I have always loved the Serenity Prayer for its humility and wisdom. Read this and consider that it was written before the nuclear age, before the digital age, and before the Internet.

God grant me the serenity
to accept the things I cannot change;
courage to change the things I can;
and wisdom to know the difference.

Can you think of how different our reaction today should be to this challenge than the people who originally heard it in the 1930s? With digital altruism and e-volunteers, the injustices that we can change are unlimited and the list of things we cannot change are disappearing.

Someone hearing this in the 1930s—when it was first recorded—would know that there are a lot of tragedies "they cannot change." That just isn't the case anymore. That feeling we all have of wanting—even needing—to do more, that feeling is our brains telling us that we need "the courage to change the things we can change." Digital altruism and e-volunteering really does require courage. It's courage not technology keeping us from making the world better.

Jobs and Freedom

What most gets me excited about virtual volunteering and altruism in the digital age is providing people with jobs and freedom. People need jobs. People want a job more than anything else. Sometimes there are huge obstacles between the job a person wants and his or her ability to secure that job and keep it. We need to be part of the solution not part of the obstacles.

In America, every night over 500,000 Americans are completely homeless. Every year over 2,000,000 Americans spend time among the homeless and cannot find a home to sleep in. Every year America has over 30,000,000 Americans among the working poor, working hard for part-time wages serving eggs, stocking shoes, and moving boxes. Every year, America has over 40,000,000 Americans living in poverty.

I've got a lot of experience among the people suffering in these categories. Don't let a couple bad experiences with individuals make you apathetic to the rest. I know that apart from God's grace in my own life I would be among them. The good news is that they want to get out—but they do face obstacles both in themselves and in society.

Jobs are not the first step, but solid employment is the crucial step alongside having a support system. Life costs money. When someone who wants a job and wants to work has an income and a support system, then he or she can be a bigger part of other people's support systems.

In my experience, it is hard to know what percent of people living in poverty want to do what it takes to have long term employment. Most are under so much oppression—oppression which others are oblivious to and even causing—that I don't think we're in a position to judge their dedication to getting a good job and getting back on their feet. We are in a position to judge ourselves and the jobs we make available. Are we doing all we can?

Good News: Potential

When we look at the American economy, we are not doing well enough at providing jobs and work environments conducive to what charity is all about. Do you really think charities want to stay in business? Maybe some do! But our goal is to see people independent not dependent. Human altruism is a biological drive to strengthen our neighbors so that they can strengthen us when needed. This has been happening and is still happening.

The logical next step is to bring virtual volunteering full force into nonprofits and businesses to provide more jobs. You cannot pay someone a wage if they are not helping create value and profits. Otherwise it's called charity not employment. Do you know how hard it is to give someone $30,000 out of your bank account?

A top notch surgeon making $400,000 annually cannot give $30,000 to a homeless family without feeling some significant financial pain since after tithing, federal taxes, state taxes, and property taxes her take home pay is about $190,000. The surgeon's $30,000 gift would be like a school teacher giving $4,000 to a local food bank. It's hard to give so much money and not feel the financial pain. However, a local muffler and brake repair center employees 15 people who all make $40,000, $60,000, and even $80,000 a year. We need small business owners!

Doesn't it make sense to virtually volunteer to help an honest business owner employ a few more people at her business and pay them each $40,000? It's not crooked. It's not unfair. It's common sense in the digital age to make the economy stronger and to help provide good jobs. The economy is not a zero-sum game. You can make the world better without making it worse for others! There is also power in decentralization

E-volunteer Revolution

and local decision making. You have very little say what happens to your taxes sent to the federal government, but you control your virtual volunteering. If you want to make an impact, seek to help free others to make an impact. Make sure your valuable time is valued by those who ask for help—and valued by yourself as well.

Chapter Seven
Volunteers Are Healthier

Volunteering improves your well-being by improving your mental and social health. If you don't have time to volunteer in person then e-volunteer. Your health and the health of others depends on it.

People who volunteer 5-8 hours a month have a longer life expectancy and a fuller life. Volunteering improves your well-being by improving your mental and social health. If you don't have time to volunteer in person then e-volunteer. Your health and the health of others depends on it. There isn't any more to say on that, but let's examine how to maximize all the extra years which e-volunteering will add to your life!

Compassion Is Critical

Compassion is a crucial part of life. Life without compassion isn't life. When people embrace their compassionate side, they improve their own lives and the lives of others. The key is to be compassionate with love not resentment when serving others. Selflessness is part of compassion. Selfishness and self-serving motives disguised as compassion or helping others harm everyone involved. Use the Internet and virtual volunteering to make sure that your time spent being compassionate about your passions is well spent not wasted.

The enemy of compassion is not the self but fear of others. Fear is natural to humans just as compassion is natural to humans. The easiest pathway for humans is to relieve personal fears, and people will always be tempted to take the route that lessens instead of increases fears. Most of the focus on improved quality of life is the removal of fears and/or the addition of joys. However, being courageous and compassionate in the face of fear leads to a separate joy of its own. Don't let fear keep you from helping the abandoned peoples. Don't let fear keep you from asking others to help you help others. Help your compassionate side be greater than your fearful side.

Deciding on the Maximum Difference

Everyone wants to make a difference in the world—at least when they're younger and more idealistic. Realism sets in pretty quickly—seems like most idealists are still single or don't have a lot of people dependent on them. However, that doesn't really matter since virtual volunteering can help everyone make a global contribution that far outweighs anything possible before computers and the Internet.

Volunteers Are Healthier

It's amazing that so many people are ignoring virtual volunteering in their life decisions. Do you know what are the main two pathways people are taking to change the world? On the one hand people are leaving the private sector to spend 80,000 hours of their lives working for nonprofits in order to make what they believe is the "maximum difference." On the other hand, people are leaving nonprofit careers to spend 80,000 hours of their lives working at their maximum potential salary while living low-budget lifestyles in order to donate individually over $200,000 annually to their favorite charities. Both are pretty good paths, and proponents of both have strong voices on college campuses pushing students toward those two options. You can either spend your whole life working at a nonprofit or you can spend your whole life as a banker and use your excess income to support three or more fulltime nonprofit employees.

Those options are great—but they were better before computers and the Internet. In fact, virtual volunteering can augment these two options instead of replacing them. Too many people are still thinking like the digital age doesn't exist. The sooner e-volunteering becomes commonplace the better—for everyone.

Three Necessities

Purpose, freedom, and mastery are known as the three keys to life. Each one is important and needs the others. Humans have to have a purpose. You can't function is life is meaningless. Humans also have to have freedom and autonomy. You need freedom, and being able to decide for yourself is key to a fulfilling life. Finally, humans need mastery or career capital. Humans get good at whatever they practice. Practice with determination and you'll get better faster. Having a skill that is in demand and a skill that you've developed has its own special magic for the psyche. Purpose. Freedom. Mastery. When you have all three of these in your life, at your job, in your family, and/or as part of your mission, you'll be happy.

The digital age increased people's ability to have all these. Virtual volunteering also increases everyone's ability to have these three. You need purpose, freedom, and mastery. E-volunteering allows you to have all three. You might never get purpose and freedom at your day job; you can have purpose and freedom when you virtually volunteer for a passion that lines up with your purpose. Virtual volunteering isn't imprisoned in the system of "management" and "administration" in businesses which can feel so unnaturally imprisoning. You can have your autonomy back to make your own contribution and learn other skills. Your employer may never pay you to master a skill you're desperate to master. You can master that skill as a virtual volunteer and also make a huge impact improving people's lives. Perhaps in the future you'll get paid for that skill too! Don't live another day without purpose, freedom, and mastery. E-volunteering can guarantee you get all three.

Maslow's Hierarchy

A lot of people point to Maslow's hierarchy as necessary for the individual's overall health. It starts with the basics of physical needs and shelter. The next most important things are social belonging and esteem. At the

Volunteers Are Healthier

top is self-actualization. (Hint: This is probably a pretty Western perspective on life. Nevertheless, I'm part of Western civilization so this applies to me—and more than likely you too!)

At the top of our personal health pyramid is the need for self-actualization, self-determination, freedom, and autonomy—these are all more or less the same thing. The digital age gives you the unprecedented opportunity to have self-actualization. You can achieve self-actualization because you have the freedom to be yourself, to be altruistic, and to have opportunity. The main things that you need for mental and social health are possible with digital altruism.

Chapter Eight
Mission and Passion

Don't ask permission to make the world better. There is no sense in asking permission from people who cannot give you permission.

Your mission in life matters a lot because you'll either spend 80,000 hours of your life working at a job or working for a charity. You get paid at either job. You'll spend 80,000 hours earning an income. You'll also spend over 100,000 hours sleeping—that's not really a choice. However, you do get over an additional 100,000 hours that are free for you to do whatever you want with—most people have kids and that reduces those 100,000 hours to about 5,000 hours, maybe a little more.

Virtually volunteering can allow someone to turn their spare 5,000 hours or 100,000 hours into a global impact that blesses millions of people! I love virtual volunteering because it's harnessing the power of the digital age for altruism. There are plenty of fulltime caregivers and people with disabilities whose lives don't have the same amount of leisure hours to make the world better. E-volunteering allows those precious few hours to multiply at an exponential and unbelievable rate.

Mission and Passion

Whether you have 5,000 or 10,000 hours at your disposal, you have to take a chunk of time to consider and solidify your mission. You're going to have between 5,000-100,000 hours to think, do, and make a difference in the world. What are you making a difference for? Most people take 10 to 30 hours to consider their mission. They repeat this 10 to 30-hour self-examination every two to five years on average because people change—so does the world and technology. 10 to 30 hours sounds like a lot, but when you consider that it'll guide your other 5,000 hours to 50,000 hours then it's not much.

Take time to consider what you're passionate about. How do your passions fit into your mission in life? Are your passions part of your purpose or the purpose of humanity? I personally identify as a rescued rescuer. I know that I've been rescued, and I enjoy helping rescue others. I know from my own experience under oppression that being rescued requires being empowered. I can't stand unfairness or bullies. I also realize a lot of people are oppressed not due to maliciousness but due to negligence and ignorance. I have experienced that even people under oppression will oppress others unless they have hope. The world has incredible joys and sorrows. Sometimes it looks hopeless to me, and sometimes humanity is this most surprisingly wonderful masterpiece-in-motion. I love people. What's your purpose? Make it more specific; make it matter. Tell others.

Passion is part of Mission

In life there is mission and there is passion. Mission is connected to purpose. Mission is part of your identity—and your life's mission can change. Your passions can be multiple. You want to have laser focus for your mission and to figure out how your passions fit into your mission. Phones, computers, and tablets were redesigned and improved by a visionary with a passion for yoga that predated his passion for computers. But his passion for yoga and electronics were part of his mission to make life beautiful and simple. He built the world's biggest company and left a lasting impact on design and user experience. Mission without supplementing passions sits on the sideline; passions without an overarching mission are ineffective. You need a mission and passions. Get laser focus on your mission and passions. Allow your passions to guide how you accomplish your mission—that's what makes you unique.

Identified mission and passions is crucial for nonprofits, businesses, employees, and volunteers—virtual and physical. You shouldn't volunteer for a nonprofit, business, or individual who can't name her or his overall mission or passions. You shouldn't volunteer yourself if you can't describe your mission and passions. You will make a big impact, and your mission and passions guide you in that. You can continue to help everyone as they cross your paths, but you have to know when and where helping people gives you the maximum joy and the maximum impact.

Mission over Money

It's important to identify, believe, live, and preach your mission. Don't be afraid to tell others about your mission. People will not understand. People will disagree. People will laugh. Some people will join you. It's worth it to find those who want to join you. If they don't believe in your mission, then the only way to get people to help you is to pay them—unless they're related to you. You can force some to help and guilt some into helping, but if people don't believe in your mission, then they will demand money. People who don't care will demand your money for their time and skills.

But people who care will put their blood, sweat, and tears into the project because they believe in the mission. (Hint: this group of people is often called a startup—everyone is passionate about the same mission. They combine their different skills and passions to accomplish the goal. They're giving their lives. Nonprofits need to have the same tenacity and cohesion! We need passionate nonprofit startups.) If you're not putting in blood, sweat, and tears into your mission, then don't expect anyone else to. Explain your mission to people and find people who share your mission. Then help them use their unique passions to work toward your common mission. E-volunteering makes connecting and collaborating on a mission possible amongst millions and even billions of people!

Mission Makes You Humble and Connected

Get your mission set in your mind. When you've got something to live for, you'll be willing to ask others for help! You cannot do it alone. You shouldn't have to do it alone. Be passionate enough to ask others to help you. Be passionate enough to move on to the next person when someone declines to help. Don't let others' disinterest decrease your interest in your mission. Other people have different life circumstances and experiences—they have different missions in life. There will be ways to partner with them in the future. You need to focus on partnering with people who are ready to join now. Hold on tight to your mission and get others to join you. We're made to work together.

Stop Asking for Permission to Lead and to Act

You don't have to ask permission to make the world better. You don't have to ask permission to do good. You don't have to ask permission to lead people. Think about that! People want to do more with their lives. We're waiting for someone to encourage us, connect us with others, and help us do more. There is no one to ask permission from to lead people and make a difference—no one except from yourself. Give yourself permission to be mocked if necessary—because it is necessary. Go lead people and then remember that success is succession. Train people to be leaders in their maximum capacity—and don't underestimate them. But first be a leader yourself and start training new leaders as soon as you have someone to lead. Maybe you need to lead yourself into action first.

Remember that most people cannot give you permission to do things—they have bosses who will tell them "no" (and bosses have accountants and lawyers who will tell them "no"). Take responsibility for what you believe and know needs to be done and do it! People who want to say "yes" can't. So do it, and then watch all the people who had to say "no" join you. Before you act in confidence, all you hear are "no's," but after your success, all you hear are "yes's." Don't ask permission to make the world better. There is no sense in asking permission from people who cannot give you permission in the first place.

Chapter Nine
Pitfalls and Obstacles

Remember that asking for help and making a difference takes courage. Remember that courage is rare. Be courageous for the ones you want to help.

Bringing volunteering into the digital age has some pitfalls to look out for. Life is not as simple now that the whole world is connected. Globalization and digitalization of economies, businesses, ideas, altruism, and nonprofits is causing a lot of disruption and change—shifts that will last. The world today is really different from the past—when everyone around you had pretty similar beliefs and your social circle was limited by physical travel. Human biology isn't made for the digital age or the space age—but we don't really have a choice! When the human body is in space orbiting the earth, the body thinks it is dying—below the astronauts' extremely red and puffy faces, their bodies are in full blown panic mode! Our blood vessels and immune systems were not made for space travel. The future requires adjustments.

It's the same with living in the digital age. Social media companies use "follower" counts and "like" counts as a dopamine-increasing addiction that the human body can't naturally handle. It's just like how refined sugar and chocolate are not natural—they have an effect on the human brain that the brain is not prepared to cope with. So let's take a look at pitfalls—both old and new—that can devastate your potential to push volunteering into the digital age.

Going Hat in Hand Hurts

Asking for help is painful in American culture. It goes against our cultural bedrocks: self-determination and self-reliance. People who ask for help or are dependent on help are ridiculed as "losers." You're not a loser. Were America's great scientists losers when they asked the president for help to make sure America was the first nation to enter the atomic age? Was the president a loser when he asked Americans to help work together to send a human to the moon? Martin Luther King Jr. was not a loser when he asked people to help share his dream.

If you have a mission, you can ask for help and collaboration.

Americans do not like to ask for help. However, Americans do like to provide help. One of the keys is being aware of how hard it is to ask for help. Just ask. Stop dreaming of what you want to do and ask for help. Asking for help might be the hardest thing you ever do—realize that is because your culture has programmed you to work alone and be self-reliant. Being self-reliant was necessary for the pilgrims and pioneers, but amazingly they simultaneously modeled collaboration. You have to take responsibility just like they did—you have to show that you're working harder than anyone you ask for help. So if you're working the hardest, don't be afraid to ask for help. If you're not working the hardest, you can still give yourself the grace to ask for help—and with the ease of virtual volunteering helping can be easy! Remember that asking for help and making a difference takes courage. Remember that courage is rare. Be courageous for the ones you want to help.

Avoid Risk Aversion

It's pretty simple why most people don't take risks: humans are social creatures and don't like doing something no one else is doing. For most people, going against the grain doesn't feel safe. When you believe and act on something that other people don't believe and don't support, it's tough. It's lonely. You have to walk a lonely road of mentally believing you are right and others are wrong. It's a physically lonely road. People enjoy hanging out with people who are like them. If no one is like you, then they won't enjoy spending time with you, and you won't enjoy spending time with them—you'll have to pretend to be someone you're not just to "fit in."

Here's the worst part about the lonely road of being different and doing what you believe needs to be done. You always know in the back of your mind that you might be wrong. Being alone was bad enough. Being both alone and wrong feels like the worst life possible—it is the worst-case scenario to many, and that's why they never do anything. But doing nothing is the worst life possible. Take hold of your mission, start doing, start leading, start improving the world! When you're a doer you can make adjustments as you learn about which parts of your mission are good and bad. Dreamers have to become doers to share the dream with others. Don't be afraid.

Reality and Fantasy

Understanding the difference between reality and fantasy is important for doers for two reasons: you detach yourself from fantasy regarding your own abilities and the abilities of others. This is crucial. Only by fighting to make your dream a reality will you see how much was truly fantasy and how much was actually able to get done. When you realize that the idea in your mind is not possible to create, then you'll also have grace for others. You cannot delegate or rely on virtual volunteers if you have an incorrect fantasy of what the outcome will be. The product is never exactly like the idea. The reality never matches with the dream. If you can't understand this, then e-volunteers will never be able to help you. What you might think can become reality might be impossible. You can either stay removed from the tangible world and keep your infatuation with your fantasies, or you can make the real world better by allowing virtual volunteers to bring your dream into reality to the best extent possible.

Know and Communicate the Need

Nonprofits have to know what is possible to accomplish, and nonprofits must communicate exactly what they need. This requires nonprofits to know enough about technology to ask for a specific need. It helps further for nonprofits to know enough about technology to make sure the specific request will make the maximum impact. The more nonprofits know about the tech opportunities and tools available, the more nonprofits can ask for each step needed in a complex project—and fill the slots with e-volunteer hours. No one likes having their time wasted at work or when volunteering.

Clearly Communicate the Need

Clear communication is needed everywhere, especially in virtual volunteering. Take the time to make sure you can communicate the specific need and specific expectations of the virtual volunteer. This is part of project management and requires the development of good processes and standard operating procedures. Develop your own manual of operations for how your nonprofit, small business, or startup identifies, tracks, and requests specific needs for virtual volunteers. Good organization and communication is helpful for everyone. The companies and nonprofits who have these attract good employees, donors, clients, partners, and e-volunteers.

Allow for Pivots and Goodness

When you do the things you love and/or things that need to be done, it makes the world better. But you can't control it. Good things take on a life of their own.

Pitfalls and Obstacles

You might find one aspect of what you do takes off and you need to focus on that. You can share your needs, get things started, connect with people, and start seeing so many people benefit from the impact you're making. Some activities will make bigger ripples for good than others. Some partners and volunteers will really want to focus on something that enlivens their passions or forms in them a brand new mission. Let them. Don't hoard people who start pursing a mission and passions that become different than yours. There are other virtual volunteers still to be found. Your future best friend might be right around the corner.

Allow goodness to flourish. When you create a positive impact, that impact is not under your control—be okay with that. Humans are made for altruism and helping one another—not fighting over resources. You can't take all the credit for rekindling altruism in others so you can't take all the responsibility to oversee the growth of their altruism. Goodness and love involve an aspect of self-sacrifice which seems to me to be unnatural to the younger human but very much natural to the older wiser human—this is a mystery to me that is perhaps related to decades of shifting memory and experiences less than innate biology: altruism and survival. Although I will add that I see people of any age young or old who have been oppressed but rescued from that oppression to be the wisest, bravest, and most self-sacrificing. I have seen many who are old, never been oppressed—we call this privileged or ignorant—lack wisdom entirely despite their attainment of power and leadership. Let's focus on the good though! Let goodness spread and relinquish control. As mentioned, success is succession, and goodness has a life of its own that when flourishing gives hope and joy on both small and global stages. I hope the altruism in the digital age can keep the scales of the future tilted toward courage, justice, mercy, kindness, and joy.

Chapter Ten
Just for Nonprofits and Small Businesses

You didn't know you signed up to spend your time seeking donations instead of making the world better full time—that's where the virtual volunteers can be so helpful!

Due to underfunding, understaffing, and constant fires to put out, nonprofits and small businesses feel more like castles under siege than a liberating army on the march.

Anything that you can pay for can also be done for free by a virtual volunteer if… you can show the specific need, do everything except what you cannot do, and find the right person with the matched skill and passion. Altruism in the digital age can spread goodness at a dizzying rate. Align your nonprofit with the best practices for making an impact and doing as much for virtual volunteers as they do for you!

Get Organized

Due to underfunding, understaffing, and constant fires to put out, nonprofits and small businesses feel more like castles under siege than a liberating army on the march. Remember your main mission and core competency. Is your mission statement up to date? Does your mission statement help you recruit and inspire volunteers? Have you been living out your mission statement? This is easy to figure out. How would the people you help and your volunteers describe your nonprofit or small business? If that description resembles your mission statement, congratulations! You're one of the few. There are few because it is hard to have a great mission statement that keeps everyone focused!

Have you been able to focus on your core competency? You need to be able to focus on what you do best as a nonprofit or small business. Did you know that most nonprofits have to spend an incredible amount of work hours, mental hours, stress, and volunteers on their big annual fundraiser? Wow! You didn't know you signed up to spend your time seeking donations instead of making the world better full time—that's where the virtual volunteers can be so helpful! Focus on getting things in order at your nonprofit. Virtual volunteers can help with expensive tasks, but remember that virtual volunteers are demanding by nature. They work at a high salary at their jobs, and they expect the nonprofits they help to be organized and true to their mission. Besides, if you're not organized, you're not ready for the capabilities virtual volunteers will usher into your endeavors.

Low-Hanging Fruit

Most of the digital services that nonprofits and small businesses are paying for are no longer needed or are available at a fraction of the cost. The logos, websites, online maintenance, graphics, phone services, virtual assistants, file storage, accounting software, CRM, etc. are no longer so expensive. You can start saving so much money today! Almost all these services and products are 90% less than when you signed up for them. Most are free! That's right. There are free versions for almost everything a nonprofit or small business needs. These products and services no longer have to be custom made. They all fit into corporations' freemium models that provide you the basic service for free hoping that you will upgrade. Out of twenty services, you might need to upgrade one or two. Why pay for full price for so many things that are free now? Why go without so many helpful things that are free?

Nonprofits and small businesses go without these things that are free because they don't want to pay someone to fix the mess they're in with an old, over-priced website on an old, over-priced web-hosting company, etc. You can hire a virtual volunteer to fix that. It's easy. List what you need and request it. Believe it or not, people want to help you! Ask e-volunteers for help. Virtual volunteers will break down the doors of your nonprofit if necessary to help save you $100-200 per month—up to $2,400 annually for just two hours of effort. Just think of how much you can save with forty hours of e-volunteer help instead of only two hours! When you add up the savings across the board it could be $2,000-$4,000 per month that e-volunteers can save you. This year you could have an extra $40,000 or more in your budget!

Another fear of nonprofits and small businesses is that all the helpful free services and apps available will become unmanageable if you're not ready for them all. Well, that's where virtual volunteers come in handy—not in managing your mess, but in helping you set up good processes for keeping things organized. There are really helpful free software applications that help manage your other software applications. All are free. All are secure. All are offered by businesses with market valuations well over $2,000,000,000—that's two billion dollars! Why aren't you taking advantage of virtual volunteers and billion-dollar companies' apps that can help you reduce the amount you are already spending? If you can't identify where your nonprofit is bleeding money, then print off your monthly expenditures and request a virtual volunteer who can examine where you can begin saving money immediately!

New Project with No Money Down

Have you been dreaming of doing something but simply don't have the budget? Do you not have the time or the money? You can still make that dream come alive. Recruit virtual volunteers! Did you know that virtual volunteers can sell tickets online for an event that hasn't happened yet? Did you know virtual volunteers can sell a product online before that product has been created, prototyped, or manufactured? Did you know that virtual volunteers can easily gauge interest in a new project, fundraiser, product, or idea before you commit additional resources or e-volunteer hours to it? There is more available than what you're aware of.

Upset Someone

Let's get back to the mission. Who are you upsetting? You can't make the world better without upsetting someone. There are status quos at every level of national life from the top world aid organizations down to your relationships with your family and spouse. You can't make a difference without upsetting the status quo. Usually it will upset a person or more too. It's human nature to not want to upset people, but you know when it's necessary. Do it with kindness, courage, and compassion, but do it.

Goals Erase Doubt

Doubt will halt your dreams. But direction will compel you toward your dreams. The first steps are setting your mission—your purpose in life—and identifying how your passions fit into your mission. The following step is goal setting. Set your goals and deadlines. Goals need to be specific and measurable but also much more. Write down who can help you, what you need to learn, and what institutions can help you. Be specific and write down when the steps must be done. When you're done writing out your goals with the requisite steps, start by accomplishing step one.

Remember that goals require the knowledge of steps and also the direction of an ending. If there is no light at the end of the tunnel, then it's hard to know which direction to go. Make the steps attainable or else you'll burnout without ever feeling a boost or confidence. You can change the world—start with yourself and your goal setting. Continue by impacting those around you. When you're confident in you mission, when you're set in your approach, and when you're organized in your efforts, you will be emboldened to appeal to e-volunteers to join you in making a difference.

Reassess and Regroup

Part of being in a nonprofit or small business seems to be feeling behind or overwhelmed. You might need to take a week to regroup—but you feel you can't take a week off. The reason most feel they can't take a week off is because they won't let a single person in need be sacrificed by their absence. I love this and support this. You can't take a week off to go and reassess your mission and core competency if that means a person goes without shelter, a kid doesn't learn to read, or a safe place isn't available. If you're like that then you're like me and a lot of other passionate people—we can't sacrifice the person in front of us to save the thousands that are not in front of us. This is hard to live with—for me at least! We all know that over 500,000 Americans are homeless every night, but those are statistics. When it comes to your local nonprofit or small business, every person matters. That's the incredible thing about people—you just can't sacrifice a single one. If you do sacrifice someone then you lose your mooring and feel even more adrift in an unloving world. We just can't sacrifice individuals.

So here's who you sacrifice—yourself. You are already sacrificing yourself, and now it's time to take that week to reassess and regroup on your mission, core competency, and how you're organizing your efforts. You can't leave those in need. So you must make up the time with later nights and earlier mornings—for just a week or two. It sounds like so much, especially on top of your overwhelming schedule. But you have to do this. There is too much potential to lessen your load and make a bigger impact. You must reassess and regroup if you're going to bring your nonprofit, small business, or startup fully into the digital age. Virtual volunteers can help, want to help, and need to help. Your mission and its success impacts everyone involved.

Ask for Emotional Support

Here's the greatest thing about emotional support. Everyone needs it! You're not a loser because you feel alone or need someone to encourage you. Everyone needs encouragement. Most successful people seek success because of their insecurities! Wouldn't it be great for the most successful but insecure people instead to have high self-worth and use their abilities to bless others more! You need emotional encouragement—everyone does. Don't be afraid to acknowledge that to others and to provide opportunities for people to encourage you. Emotional support is one of the foundational needs of humans. Consider encouragement to be one of the most basic parts of altruism and belonging.

Here's the second greatest thing about emotional support. Everyone can provide it, especially with the advent of the digital age, the Internet, and social networks. Perhaps I should have said the second greatest thing is that emotional support is the easiest virtual volunteer skill! It's an entry-level e-volunteer skill! Anyone who can send an email, send a text, or access a social network is a person who can send you encouragements and keep you going toward your goals. You might take the simple step of stating the hard task you're taking on that morning and ask for people's "prayers, thoughts, and/or encouragements"—or let them choose which they want to send! If you need to be more specific, be more specific. If you need to create an emotional support group for yourself, do it. Ask for emotional support. Other people are counting on you to carry out your mission; count on other people—maybe you are part of their mission?

Hope for the Hopeless

It's a known fact that when people are faced with horrible options in life that they make the worst decisions. When faced with horrible options, humans tend to choose the most impossible option with no hope of success. This is awful, and it reflects a person's knowledge of their impossible situation. In the face of hopelessness, people latch onto an irrational hope that a crazy risk will save them—it can't and never does. However, this latching onto a 0.0001% crazy risk is a relic of the past. It's the irrationality of believing your only hope is winning the lottery. The truth is that a new digital lottery has already been won by everyone who takes ahold of e-volunteering to bring altruism into the digital age. The things that you want to accomplish for yourself and for others are all possible by calling upon virtual volunteers. It's really unprecedented and incredible—and who is it that normally faces the most impossible odds and obstacles: the dreamers, disruptors, and rescuers of the oppressed. Digital altruism is for all of us.

The Secret to Asking for Help

I could go into all the reasons that some are able to ask for help and some are not, but I will not because it is sure to offend most people if I pull back the veil too much. There is a lot of science behind the social, cultural, and biological reasons that some can ask for help but many feel unable to ask for help. Nevertheless, it is hard to ask for help. Here is a secret I've found from my own experience. Don't ask others to help you. Instead, ask others to help you help others. Put the focus on the others that you are helping. You need help in order to help others. Focus on that, and make sure you are actually helping others. This secret will change everything for you, and it will empower you and keep you accountable to improve the world.

Clear Communication

It is worth reiterating one more time that communication must be clear and honest. You must communicate your expectations. You must communicate clearly. You must reassess your ability to communicate every time there is a breakdown in communication—it is your responsibility to be in charge of the communication not the e-volunteers' responsibility. You must make your communication concise not overly redundant, but you also make your communication foolproof. That is not an easy combination to achieve—simple yet foolproof! That is why clear communication is so valuable. Become better and better at it. You cannot organize the endless potential of e-volunteers without clear communication.

E-volunteer Too

You have the skills to e-volunteer too. You have been able to lead a project, team, nonprofit, small business, or startup! You can already help people with consulting and networking. People have dreams to do what you do, and you have the experience to guide them to get where you are—with their mission. You can also develop new skills by e-volunteering for other nonprofits or businesses. It doesn't make much sense for everyone to be volunteering or e-volunteering except for the directors of nonprofits. Okay, actually it does make sense because nonprofit directors are underpaid and work unpaid overtime hours—which itself is pretty much volunteering. However, don't underestimate the joy and knowledge you'll get from experiencing e-volunteering for others. When you're no longer in charge, you'll get a different perspective on e-volunteering! You might not have time to volunteer in person for others, but you do have time to e-volunteer. It's also a great opportunity to

Just for Nonprofits and Small Businesses

meet others who share your mission and need your skillset! Yes, e-volunteering for someone else takes time that you do not have, and adding an additional task to your schedule seems stress-inducing. However, e-volunteering will make you feel better, feel less stressed, and give you more time not less because of how much better you'll feel physically and socially!

Chapter Eleven
Just for E-volunteers

You have a mission and a skill, and the organization that needs your help has a mission and a need for a skill. Make sure your mission and skill align. The more alignment the better.

I'm excited you are considering how to maximize your volunteering even more! Remember that relationships are just as important as seeking to maximize your time. You can develop great relationships volunteering in person or collaborating online as an e-volunteer. You never know where a relationship or altruistic endeavor will take you!

I want to thank you for all the virtual volunteering you are already doing. You probably don't realize the impact you are already having when you help friends, family, and co-workers with your skills that you donate online, over the phone, or through text messages. Even extreme introverts have over 10,000 interactions during their lives. All our interactions can be positive or negative on others! I'm glad you want to make the maximum positive impact while still having fun!

Lastly, I want to encourage you to say "No" because sometimes you must say "No." You have skills and passions yourself. Use every means available to you to help those who need you the most. Don't forget to help those who desperately need your help but don't share all your passions—altruism sometimes has to put people over passions. However, beware the wolves who will want to use you for your skills and not your passions. If you don't feel like it's true collaboration or in your mutual-interests to help, say "No."

Slacktivism

Activism is important to make a difference; the biggest obstacle to activism is time commitment. That's why people who comment on social media or talk about making a difference but never act are ridiculed as slacktivists. It's basically a "slacker activist." They're the people who say a lot but don't do a lot. They're trapped in slacktivism—probably because they don't have enough time not because they don't want to help. It's not their fault they don't have more time to help.

The good news is that e-volunteering is easy. People get trapped in slacktivism because making simple comments on social media is so easy. It's easy to do something you're good at. That's pretty much the definition of easy. It isn't easy for a baby to learn to walk, but most adults consider the ability to walk to be very easy or second nature. Virtually volunteering with a skill that you're good at is one of the easiest ways to make a difference. So try to make sure you comment less and do more—by aligning your passions with your skills.

Do More and Record More

When you're e-volunteering for others, you're making a difference and blazing a trail. Share the processes which you discover work well for you. You're really just working remotely online for someone who can't afford to pay you or for someone you won't allow to pay you. First use the skills you've learned working virtually, but, as you get better at making your skill available to nonprofits and small businesses, record the standard operating procedures that help you e-volunteer. Recording the processes will help the organizations and other e-volunteers. Being a difference maker has many ways to bless people.

Match Mission and Skill

This is the most important part of virtually volunteering—it's where the efficiency and joy is maximized. You have a mission and a skill, and the organization that needs your help has a mission and a need for a skill. Make sure your mission and skill align. The more alignment the better. If a nonprofit that helps sell jewelry made in India needs help with a composite product plug-in built on WooCommerce built on WordPress, then the more you align with their mission and need the better. Even if you're a computer programmer, if you don't have knowledge of how WooCommerce works, then you might not be the best person. Allow someone better aligned to help. It's necessary. Allow the Internet and the networking power of the digital age to make the best match. Focus on

Just for E-volunteers

helping how you can with the things you care about.

Mission and Passion

Have you taken the time to consider what your main mission is and what things you are passionate about? Remember these things can change and for some personalities these things stay in more or less a constant state of flux. Passions can be fads or things that are trendy—that's okay. Your passions will never go against your mission! If it seems your passions are going against your mission, then either one of them is wrong or you're not being creative enough.

On the one hand, perhaps your differing mission and passions are telling you that you've made a mistake in analyzing your mission or passions. Perhaps it's time for a shift—these can be scary but you have to do it because our identities and allegiances will shift in life.

On the other hand of this whole scenario, maybe reconciling your mission and passions requires some creativity and courage! Humans haven't always been space travelers, and humans haven't always been able to talk to someone 3,000 miles away in real time. Perhaps you are about to invent a new combination or something totally new. Either way, if you do not identify your mission in life and your passions then you won't have a foundation to build on.

(Tip: it is terrifying to have a mission that your family or religion says is not okay. But you have that taboo passion for a reason, and you won't be able to be you unless you live out your mission according to your passions. You will be lonely whether you act or not on your mission. When you do begin to act, you will meet more people and be less lonely. Family always comes around too. The good news is that you'll be happier, and if you can spread joy, then that will open more hearts.)

You Have Skills

No one has zero skills. Everyone can encourage others. Everyone has the freedom to share and to help advertise someone's nonprofit organization, events, or products. Many people have more skills such as copy editing, proofreading, and content creation—ranging from amateur to professional abilities. Fortunately, all levels of skills are needed and helpful. Even better, when you e-volunteer in these areas, you become even better.

There is another set of skills that are software application-based. These are skills that everyone has using specific software applications that are mobile-based or browser-based. Companies and nonprofit organizations use between 20 and 80 applications each. Perhaps you have skills in one of these. Maybe you don't have skills in an online accounting application but you have talent for Instagram. There are a lot of nonprofits and companies fumbling with Instagram who need you to take 10 minutes to look at their pictures from the past six months and tell them which ones—maybe up to half—to delete! Perhaps you can give them more input on keywords to hashtag or do a basic audit of their social media strategy for that platform. They might be scheduling posts for 4am instead of more engaged times of the day! Help by changing a few things. Help by donating a few e-volunteer hours here and there. You can make a big impact for a mission you care about. The good news is that there are hundreds of business-

Just for E-volunteers

applicable apps that potential e-volunteers are using during their leisure time. You can put these skills to work for people you want to help.

Next, let's further examine skills and software applications you use for your job. There are many expensive software suites that you and your co-workers are using—and nonprofits and small businesses use these too! You can really help someone out with a little free consultation or free help when you have the time. Nonprofits and small businesses usually don't have the capital to dedicate a salary to every specific role. Maybe you work for a larger company or a niche small business and you have incredible skills on important e-commerce syncing software such as Unify, Stitch Labs, or TradeGecko. Yikes! A small business or nonprofit could waste $30,000 or lose $200,000 in missed opportunity if you don't provide a 30-minute consultation or an hour of e-volunteering. If your skill and mission match the small business or nonprofit organization's need, then there isn't a better use for your volunteer time than to help e-volunteer! This is how you make a $200,000 impact in less than an hour! Volunteering really does pay!

There are also a lot of mobile-based applications that you use for your job that make your work ten times more efficient and make your company ten times more revenue. All these apps are available to nonprofits and small businesses—usually on freemium plans too. A little e-volunteering can help a nonprofit organization begin to use five to ten to twenty apps that will revolutionize the nonprofit's ability to make a huge positive impact. Make sure the nonprofit organization starts with one app at a time of course!

Lastly, there are programmers and coders (hint: coding is just the last stage of programming) who have been e-volunteering since they created the Internet. Did you know that e-volunteers are always open sourcing things that people need? In all honesty, the e-volunteer opportunities mentioned in the prior paragraphs are just a way of allowing non-coders the ability to enter virtual volunteering and digital altruism—although some see the open-source revolution as anarchy.

Nevertheless, e-volunteering really began with programmers making free and open source software (FOSS). Here's some examples you know and some you might not know but have benefitted from: WordPress, GNU, Linux kernel, Mozilla Firefox, and Open Office. Nonprofits and small businesses can benefit from the many open source software available.

Coders can also help e-volunteer as tech advisors to read the code that nonprofits are paying for—this is probably the most basic high-level e-volunteer aid besides consulting. Of course the right consulting is invaluable in and of itself. E-volunteers can also help nonprofits and small businesses design the UX and figure out the necessary technology stack for a project. In my experience, it is much easier for e-volunteers to help make an internal business app or an app that is a minimum viable product (MVP). Of course e-volunteers are free to collaborate on 200-hour projects that would normally be billable for $40,000 to $100,000 dollars on the open market. Such an undertaking is indeed possible with the right mission and communication. This is the power of altruism in the digital age, but I would suggest starting small and focusing on one-off tasks. However, if the mission and skills align, and if there are enough collaborators, then something fantastic could arrive sooner rather than later—that was the case with many of the world-changing, open-source projects listed above.

Just for E-volunteers

Learn Skills

Another great opportunity just barely mentioned to this point is the ability to learn a new skill. When you e-volunteer, you can learn something new—just don't trick the nonprofit or business into thinking you have 20 years of experience. Do your best. There isn't any risk. You're volunteering. You aren't getting paid, right? Well, even though you're not actually getting paid, if you cannot fulfill the request then communicate clearly and quickly that you cannot accomplish it—there are many e-volunteers who can step in at a moment's notice. You didn't get paid and didn't plan on getting paid—you tried and that was enough. Money didn't matter to you. Fortunately, you still learned more about the skill, communication, and the mission you share with another person or group of people. If you "fail forward" then you'll be ready to e-volunteer again soon. Failure is only bad if it causes you to stop e-volunteering and stop maximizing your difference.

People pay to go to college—usually $20,000 to $50,000 annually. College students pay money whether they develop the skills or not! You didn't pay any money to e-volunteer, and you actually volunteered for a real nonprofit or business. (Tip: it's important to make sure not only to match your skill and your mission with a nonprofit or business but to also make sure they won't waste your time. Usually a nonprofit or small business are less likely to waste your time than a solo-operation or an entrepreneur. You can help an individual, but it helps if they have a history of getting things done and making a difference. The same applies to nonprofits and businesses. Check their track record.) If you want to learn a new skill or perfect a skill, then start to look for e-volunteer postings for that need. That specific skill might become a promotion at work or spur you to create your own business or nonprofit in the future!

Feel Better

The cure for a lot of depression is to do more for others. Altruism is social. Don't be selfish; don't look for how you benefit. Finding a virtual skill you can donate and aligning with your mission is important because you produce better work, enjoy it more, and actually see it through to completion. Even if you are not suffering from depression, doing good for others will always make you feel better. It's also the right thing to do. Now it's easier than ever too! When you e-volunteer for causes you care about, you feel good and make others feel good. You are helping open doors for others as well as for yourself. Since it can be hard to keep the altruism the focus instead of yourself, I recommend keeping the focus on finding an e-volunteer task that is the best fit for your own mission and skills. You'll be surprised how big of an impact you have and how much better it makes you feel!

Stay Connected to People Too

In life there will be some people who treat you like an object and some people who treat you like a person. We all want to be treated like a person not an object. Keep an awareness of how you treat others too! When you only care about someone because your missions and passions align, then you're treating that person more like an object than a person. Altruism is about people not objects. Let your aligned interests be what connect you as two people—not objects. You will burnout if you view people as objects, and your altruistic acts might be derived from your own insecurity more than your love for others.

Matching your specific skill is still important. You will make such an enormous impact when you use your skillset wisely for the e-volunteer revolution. Remember to get to know the people you're helping and who share your mission. Keep some boundaries, but remember that they're people not objects. Make sure they don't treat you like an object either! The e-volunteer revolution is about maximizing our resources via technology to help people. The people are the goal. If you start to feel like an object or see other people as an object, then take some time to refocus on altruism's love for the individual human with fears, sorrows, joys, and loves of her or his own!

Don't Be a Scrooge

Some people just don't feel freedom to give away their time or their skills to those in need. This isn't a problem with the person; this is a problem with society and culture. Cultures are complex systems that guide society in who and how to help. Society is not valuing digital altruism the same way society values medical altruism. Society is wrongly telling people that medical skills are an inherent good for society while skills like photography and computer programming are in a lower category than medical expertise. Well, digital altruism

E-volunteer Revolution

says that your skills are crucial to making the world better. Everyone needs to volunteer five to eight hours a month for their own social and physical health—not to mention the world's health.

Consider that many people are photographers and have photo editing skills. They more often than not choose when to volunteer their professional skills instead of being bombarded with requests to volunteer. In fact, because society has told them their skills are not vital to life, few ask them to digitally edit existing photos or to take a 10-minute photoshoot—you really can't ask a photographer to volunteer a whole day of taking photos. Of course people make requests here and there, but for the most part, photographers are not bombarded with requests from nonprofits for e-volunteer work. There is a reason for this. Everyone needs to make money, and society has the understanding that photographers need to be paid—this is true. However, everyone also needs to be able to volunteer at least 5 hours a month at what they're good at. When everyone e-volunteers together, then the world's relationships and institutions and economies are stronger. Part of e-volunteering is the photographer gets to choose when the match fits her or him not constantly declining requests from nonprofits whose missions don't interest the e-volunteer. The e-volunteer revolution is about creating a win-win for everyone. That's the beauty of digital altruism.

Let's contrast the photographer who is being asked to e-volunteer as a photo editor for a nonprofit compared to a doctor being asked to do medical missions. Society knows the doctor makes more money on average than a photographer and can afford to volunteer a week a year or every other year. However, without good photos of the doctors volunteering their medical skills, it will be harder for nonprofits and socially-minded startups to show the world the need for doctors to volunteer in impoverished local and global communities. Photography and altruism is incredibly important. E-volunteering helps make sure people's precious time and skills are not wasted.

Society's status quo perspective on the value of volunteering is wrong. Volunteering is dropping every year as people become busier and busier. It's hard to put firm dates into our schedules and to spend time driving to physical locations for volunteering. The beauty of digital altruism is that volunteering is still possible and even more possible than ever. Not only is it easier to volunteer, but it's easier to make a bigger impact volunteering. Whatever your skill is, you are in control of it and in control of your life. Be open to e-volunteering with your skill. The problem you might have encountered henceforth is not knowing your own mission and passions. It's easier to e-volunteer when you're excited about it. Stop volunteering for things you don't want to volunteer for. The world needs you to be you. The world needs your altruistic version of you.

Match Matters

Once again, focus on matching your skills with your mission when nonprofits or individuals ask for help. You must use the Internet to connect to people, and you must force your friends who need help to also use the Internet. You must enlarge the pool of opportunity for all. You will often be the only person with any digital skills that your friend knows—that's why they are asking you. Tell them to find someone who matches better on mission and skill for what they need. If you don't, then you will waste your time and their time.

Matching mission and skill is more important than matching on friendship. You will be asked repeatedly by friends to help with their passion projects. You might not have the specific skill or specific shared passion. Let them know that. Stop saying "Yes" when you need to say "No." When you say "No," mention to the requestor that there is a better way for her or him to tap into the e-volunteer community to find someone who matches better for both that mission and that specific skillset. You're the person's gateway to discovering virtual volunteering. Help them that way.

R.O.W.E.

The results only work environment is your best friend in virtual volunteering. The people you are helping often don't have the money to hire someone, and they are in need of e-volunteer help. You have to take care of yourself and make sure the people you're helping are not abusing your time or efforts. Remember that nonprofits, small businesses, and startups are often hectic and juggling too many things at once. Don't allow yourself to become one of the many things they are attempting to juggle. Help places that are under good administration and run orderly. If the communication is bad at the beginning, it is not going to improve. Remember this: bad communication at the beginning is not going to improve. Bad communication is a sign of disorganization that will make collaboration impossible. Your time and the difference you can make when utilized well means you have to say "No" when you realize it isn't going to work out.

You are being asked for a result. The result is what matters. Your job is to let them know when you can't give them the result they need from their e-volunteer, and their job is to make your e-volunteer task as easy as possible. If they are not communicating well, treating you well, or even being positive people, feel free to stop helping immediately and let them know immediately. Never stop working on a project without letting them know you have stopped. Furthermore, try your best to let them know what they did on their end that made your ability to help difficult—your constructive criticism could save their whole endeavor or organization! Don't let others take advantage of you. Your own mission, passions, and specific skills are too important to the world to be wasted by anyone.

Match Matters more than Impact

Don't get trapped in paralysis by analysis because you are looking for the nonprofit which makes the biggest impact or stretches its money the furthest. Too many people waste time doing this instead of taking advantage of digital altruism. Nonprofits are successful or unsuccessful due to many reasons which you cannot know completely. However, you can know your own mission and skillset, and you can know when your mission and skillset aligns with an e-volunteer opportunity. If the organization can clearly communicate and not waste your time, then volunteer for them. Don't worry about the maximum impact or how solid the nonprofit is. Your matching your own mission with your own skill as an e-volunteer will begin to change the world in a unique and powerful way that is only unlocked in the digital age. The future's best nonprofits and most world-improving businesses might be the small team whose mission and needs align with you! Help them and have fun!

Volunteer for Businesses and Startups

Nonprofits are dependent on funding and donations from individuals, from businesses, from governments, from foundations, and from the products that nonprofits sell. Individuals have to make an income to be able to donate. The same is true for businesses. Governments have to tax the profits of businesses and individuals. Money or value has to be created in order to make the world better. Money or currency is just a place holder for value. When you volunteer you make the world better. You add value. When businesses sell products without manipulative marketing tactics or dishonesty they make the world better too. They add value.

You will never find a business, nonprofit, individual, or startup which acts with 100% honesty or perfection. That's part of being human. However, you will find plenty of people and entities trying to be better. Look for organizations that get things right more often than not. Look for organizations that make the world better regardless of their intentions. Be honest with your own faults and keep a positive attitude. Perhaps you can help their organizational culture while you e-volunteer for one of their desperate needs. Seek to help others make the world better. Beware of nonprofit organizations or companies that rely on marketing tactics or manipulation instead of providing real value.

Usually it is inventions, disruptive technology, new technology, and solidarity with the oppressed or immigrants that make the world better. In these areas, there are millions of nonprofits and small businesses that you can e-volunteer to help! Out of the millions of nonprofits and businesses, there are many who align with your mission and skillset. Help them make the world better. Don't worry about how you're benefitting. If you're only volunteering 5 hours a month, the benefit is in helping others altruistically for something you care about. And trust me, there will be many other benefits too! (Hint: the benefits are new skills, better skills, making a difference, living longer, finding your mission, fulfilling your purpose, connecting with people like you, etc. It is a pretty good list, and it just keeps getting better.)

Setup for Success

No one likes the co-worker who tries to make sure nothing under his oversight can function unless he is involved. In his mind, it makes him indispensable and provides job protection—no one can fire him or the wheels will come off the organization or project. That doesn't sound like altruism. That sounds like laziness or insecurity.

Always leave a place better than you found it. When you help a nonprofit, don't leave a mess behind you. Don't make yourself indispensable—that's not altruism. Make sure you help establish processes so that others can e-volunteer as well. Help nonprofits and small businesses have their online portions of their organizations manageable and organized. E-volunteering is both helping and making sure the next step in the process is clear. Others will be e-volunteering for this company or nonprofit as well. Don't make those e-volunteers' tasks harder. Make things simpler and remove the confusion.

Go from E-volunteer to Founder or Director

You have the skills to e-volunteer. You also have the skills to lead your own project, team, nonprofit, or startup! Maybe this is something you do sooner or later, but always be aware that your ideas are good and your mission matters. Join others of course, but if necessary make the transition to organizing e-volunteers who align with your mission. Take the things you've learned from e-volunteering for others and do your best to incorporate the good things into inviting others to e-volunteer with you!

Appendix A

The Best Apps and Online Tools for 2018-2019

I've narrowed down this concise list from over 2,000 possible options in about 40 categories. All 2,000 options have teams of dedicated computer programmers and anywhere from 100 to 10,000 employees. I've tried to give you the best one or two apps in each of the main categories. I used the following metrics for ranking them:

- ease of user experience
- potential for virtual volunteers
- size of user-base support community
- customer service
- integrations with other applications
- price

Remember, we suggest using a 3rd-party company for dedicated password storage, for password protection, and for use by e-volunteers. A password service adds an extra layer of trust for online work done by employees, free-lancers, and e-volunteers. You can find good password companies and project management companies at VolunteerHill.org or on the following pages.

Space for Your Notetaking or Needs:

The Best Apps and Online Tools 2018-2019

Accounting:

- Wave
- FreshBooks

With good accounting software you can save over 200 hours a year—that's five of your workweeks! You also have the ability to plug in these online applications to all your other online tools. The $6,000 version of QuickBooks cannot do that. That is why Intuit is pushing QuickBooks to be fully online compatible. I suggest going with FreshBooks if you're a nonprofit or business. If you're testing an MVP or a personal dream, then Wave is right for you!

Ads:

- Google AdWords
- Facebook Lead Ads

Marketing is not free. You have to have some ad spend at some point. Both of these platforms have the most users online. A little ad spend done the right way can make a huge difference. $300 spent on the right ads can generate $1,000,000 with of views if you have the right content and the right community sharing and talking about the ad! Even for small ads, be sure to make the ad's wording good, and e-volunteers can help with A/B testing. For content make sure it is sharable and has the right call to action.

Analytics:

- Sumo

There is nothing better than Sumo for your email collection, heat maps, and generating shares. There is a free version as well.

App Coordination:

- Zapier

There is no competition here. Zapier can connect all your apps. It allows your apps to talk to each other and work in the background. Zapier lets an action by one app, such as receiving a chat question, trigger an action in another app, such as sending an email. There is no limit to the steps, but be careful not to create a loop on accident!

Bookmarking:

- Pinboard

Tagging good items is a crucial part of the knowledge economy. You need to be able to create your own Pinterest board by using Pinboard to tag articles and pictures that are critical to processes and onboarding for your mission, projects, passions, and clients.

Calendar:

- Google Calendar

This is the best. You need to get it integrated ASAP if you haven't.

Chat:

- Intercom
- DialogFlow
- ChatFuel

Intercom is being challenged in this category by ChatFuel powered by Facebook. Dialogflow is the most powerful and powered by Google, but it is also the least aesthetic. Use Intercom if money is not an issue, but use ChatFuel if you have a heavy presence on Facebook since it integrates fully with Messenger. DialogFlow should only be pursued if AI is preferred over aesthetics and you have a long list of virtual volunteers ready to jump into AI.

Contact Lists:

- Google Contacts
- Full Contact

These powerful tools can keep track of your contacts across all platforms, and they can keep them in one place. Use the one you prefer or use both simultaneously!

Customer Relations Management:

- Pipedrive
- Hubspot

Hubspot CRM is free and sufficient for all your needs. Pipedrive currently has the advantage of being better if you do more on your mobile phone than on your laptop. Pipedrive is especially useful if you usually only have your mobile phone when with clients or collaborators.

Customer Support:

- Intercom

Intercom is the industry standard here. Intercom helps you set up a chat that will answer questions, refer inquirers to helpful articles, and fetch a human as needed.

Dashboards:

- Databox
- Cyfe

Dashboards can give you a bird's-eye view of

your organization. They'll alert you which areas to improve and which people to pat on the back. Databox is simpler. Cyfe is more complex but probably a better option.

Design:

- Canva
- Typeform

You need both of these. Canva is best for static designs like banners, images, etc. Typeform is best for dynamic uses such as questionnaires, surveys, forms, and contests. You will likely use Canva a lot more for designing anything from social media posts to book covers.

Email Campaigns:

- MailChimp

There are not any other options here. MailChimp is worth the price because they are the best at getting your sent emails past spam filters, actually opened, and achieving the call to action. MailChimp is easy to use and helps make beautiful emails which are plugged into analytics.

Events (Big):

- Eventbrite

The only way to go for big events is Eventbrite. It takes care of everything and brings legitimacy to the event. Don't forget to connect it with all your social networks too.

Events (Small and Recurring):

- Meetup

Meetup is really best for recurring events. It can build a lot of goodwill and connections for a mission or passion. Meetup is popular in the tech community. You can also achieve a lot with Facebook events.

File Sharing and Storage:

- Google Drive
- Dropbox

Both of these are essential. Use them to store everything except passwords! Seriously, don't store passwords in these.

Forms:

- Gravity Forms

This is the best tool for making forms that people want to fill out online or in person. You want people to have a good experience whether they are in your clinic or on your website.

Goal-Achieving:

- Todoist
- Wunderlist

Both of these are great for plugging in goals and tasks. They help you tackle your goals and keep track of your metrics and progress.

Keywords:

- Mention

Google's options for keyword planning can be difficult for the uninitiated. Consider using Mention for its high quality UX. Your other option is to go ahead and get familiar with Google's options for keyword planning.

Marketing:

- ClickFunnels

You have to organize your marketing and do more than advertise. ClickFunnels will help you get your customers and clients where you want them on your website. It will help you learn more about what your customers too want so that you can make sure to offer it and meet their needs.

Notetaking:

- Evernote

There isn't anything better to keep you from being overwhelmed with all your ideas.

Password Storage and Use:

- LastPass

The best option is LastPass although there are other options that are free. You need password storage so that employees as well as third-party freelancers and e-volunteers can access your many applications without having to give them your passwords, change your passwords, or fear sabotage.

Phone Sales and Follow-up:

- CallRail

You can run online and offline campaigns. You can use referrals or direct call lists. There is also an enormous potential in tracking conversion ratios for potential partners initially found in pay per click and SEO.

Photography:

- Unsplash

Free photos that you don't have to worry about licensing. Incredible. Provided by e-volunteers! There are also a lot of other online options for inexpensive high quality photos as well as free photos that are under a creative commons license.

Project Management:

- Asana
- Trello
- Basecamp

My favorite here is Asana for overall value, but Basecamp is the most used and most expensive. Give Trello and Asana a try. A small team can get by with Asana's free version.

Proposals:

- Proposify

Out of all the proposal software out there, this one will get the most customer conversions, simplify your proposal construction, and create an invoice to send your accounting software or send straight to the client.

Good proposals are part of the excellent communication needed for e-volunteering and for nonprofits and businesses in general. Clearly setting out the expectations at the beginning makes a big difference.

Recruiting and Hiring:

- Breezy HR

This will help you get the people you want and find the people who want to collaborate with you. You need the best match possible if you're going to recruit good e-volunteers and co-workers.

Scheduling:

- Calendly

There isn't anything better. You should already be using it—everyone else is. Calendly will help you get control of your schedule and be more efficient with your meetings and networking.

Signatures:

- SignNow

This works on any type of device and looks great. If your main use for signatures is a PDF or a Word Document, then consider HelloSign.

Social Media Posting:

- Buffer
- Hootsuite

These are scheduling tools that can also track analytics. They include free versions as well. Buffer is good for evergreen content, and Hootsuite is good for curated content. Almost everyone uses both simultaneously, and you should too!

Social Media Listening and Monitoring:

- Buzzsumo

You want to use this to find out what keywords and topics are trending in your niche. It gets you ahead of the competition by giving you all the insights regarding your niche product, audience, sector, strategy, and market.

Surveys:

- SurveyMonkey

Surveys are helpful when done right. The responses and analytics available can save your nonprofit or company from making major mistakes.

Teamwork:

Slack

This is the best for organizing teams and staying organized. Slack is an enormous company that can help a company of any size.

Selling Online:

- WooCommerce
- Shopify
- Etsy

WooCommerce is the best e-commerce option due to many reasons not to mention its enormous support community. If something easier is needed, then Shopify creates a basic site and has even less expensive options for integrating ecommerce onto other connected platforms. Use Shopify's most basic version if products are a minor part of your revenue stream. If you need a place to test products, consider Etsy. Magento is not on the list since it is best for large companies and scores poorly for amount of people available to provide free support. Amazon and Ebay are also great tools to use, but those would be next steps—especially Amazon. Selling at a high quantity with drop shipping and constant price adjustments is right around the corner.

If you don't have enough products to fill up an online store, then a good starting point would be using Shopify or Facebook to sell products directly to your fans and supporters through Facebook or Instragram.

Teaching Courses and Training:

- Teachable
- Thinkific

Both of these options help you construct courses and host them from your own website. You can create courses to sell and free courses for clients or collaborators. This is a great way to spread your message about your mission or give other people tips about how your process is effective. These companies will help make your message clear and keep the ratio of completed courses higher.

Web Hosting:

- SiteGround
- WPEngine

Both of these are excellent for most websites. They load quickly and have extra security. They're normally used for hosting WordPress sites. The differences in pricing reflect quality, but both of these companies are the best for their price points.

Websites:

- WordPress
- Wix
- Squarespace
- Shopify

You really need to have your website built in WordPress for both SEO and for the enormous support community. WordPress has the biggest community online and no shortage of virtual volunteers. WordPress was created by e-volunteers, and it's also the best in every category except ease of use for people with no computer skills. Dive into WordPress and do not worry about the learning curve since help is readily available. The best ecommerce option WooCommerce requires WordPress as a prerequisite. Oh, and remember to use Wordpress.org not Wordpress.com. The dot org is the awesome open source community of opportunity. The dot com version is a more expensive and lower quality predecessor to Squarespace.

Wix and Squarespace are quality options if you do not believe WordPress is right for you. If you do not go with WordPress then make sure your Wix or Squarespace site is built and maintained with a future switch to the more powerful WordPress in mind. A Wix or Squarespace site can support a large business for years if needed. If you missed the ecommerce section and mainly sell products online, you need to use WooCommerce built on a Wordpress.org platform. If WordPress is too complicated for you at your current stage, then go with Shopify.

Appendix B
E-volunteer Resources

E-volunteer.com

E-Volunteer.com helps match needs with skills. It functions like Airbnb but for virtual volunteering. In that way it's more like an e-volunteer version of Twitter. It's a way to share what you're doing instead of what you're thinking. It is made for matching skills and passions between e-volunteers and those who need help. It's a place where someone asks for help, and help appears.

TandemAI.com

Tandem AI helps humans race with machines—machine learning to be exact. Tandem AI is an AI that helps simplify and assemble a project/goal according to needs, skills, and passions. This is helpful for e-volunteers, nonprofits, entrepreneurs, businesses, and individuals. Anyone can ask Tandem AI for help starting a business, completing a project, or getting something done. Tandem AI narrows the to-do list and outsources the work to e-volunteers and to bots for completion.

The long term goal of Tandem AI is to continually replace human actions with machine learning. The end result will be more time for human creativity while AI takes care of processes, automation, and technical aspects. Tandem AI also has a project manager software that helps put projects and ideas onto an assembly line of e-volunteers and automation to bring ideas into reality.

TECDonor.com

TEC Donor is a very helpful company to get your nonprofit signed up with. TEC Donor can help coordinate and reward e-volunteers. This is important because even digital altruism benefits from a win-win, especially if the e-volunteers can get actual gifts and products after volunteering.

VolunteerHill.com

Volunteer Hill is a place for e-volunteers to list their skills and passions. It functions like Fiverr, Ebay, or Amazon. The skills have already been listed online by the e-volunteers. Skills are free or reduced by 75% to 80% for nonprofits and socially responsible businesses. The fee for services is minimal but exists to make sure that the e-volunteers' time is not wasted. The money spent goes to other nonprofits and goes to supporting crucial e-volunteer awareness. If you have an e-volunteer skill and want to list it, this is the place.

VolunteerHill.org

This is the home for e-volunteer awareness online. The site also shares best practices, e-volunteering listicles, and good stories from e-volunteers around the world. Everything that is making the world better is posted here. If you want to learn more about how you can benefit from e-volunteering, then start here.

VolunteerMatch.org

Volunteer Match is focused on in-person volunteering not e-volunteering, but usually 5% to 10% of the listings on Volunteer Match can be done remotely. However, remote volunteering is not always as simple or easy as e-volunteering.

Appendix C
The Best Social Media Platforms for 2018-2019

You have to use the big platforms to reach people, customers, and e-volunteers. Remember that when running your own life, a nonprofit, a startup, or a small business, never try to do more than you can keep track of. Start with one thing and then build up more. There is so much possible and so much that is helpful—it's too much. E-volunteers can help a lot but don't overburden yourself. Here's what you need to focus on in order of importance. For each of these you can research best practices or reach out to an e-volunteer to help navigate the best pathway for you.

These social media platforms are some of the biggest companies in the world today. They have huge numbers of people on them—people who need to know how you're making the world better through your cause. Are you looking for new customers or supporting existing customers? That will affect your choices here.

Personal Facebook Account and Facebook Group for Nonprofit or Business:

Facebook is the largest online platform. It would be a social network but because you're advertised to so heavily on there it's social media. You're advertising your mission on there too, so, yeah, it's "media."

It might be surprising, but you need to have a Facebook group first because this group provides feedback. It's tough to start with a Facebook group instead of a business page because we prefer fantasy over reality—but the tough road is the only road to success. A Facebook group connects you with people. You're creating this Facebook group for your official organization or for the main mission of your official organization. (Tip: if you're a virtual volunteer you might create a Facebook group just for your mission or skill or both as one group!) The Facebook group is an unlikely priority, but it is the most valuable thing you can do on social media to get started. Find the community, build the community, and learn from the community. The ROI for your Facebook group can do so much more for your mission and brand than anything else.

YouTube Channel:

YouTube is so important because of its SEO metrics. You need to be on Google's radar, and YouTube gives you a way to feature your mission, your expertise, and your services. Stay up to date and focus on the SEO incorporated into the platform.

Facebook Page:

In a pinch, a Facebook page can replace a website and even incorporate e-commerce. The chat and messenger functions are powerful and nearly limitless too—definitely an e-volunteer endeavor. Facebook pages are key and do not require too much upkeep. Since your Facebook page is your business or nonprofit organization's official presence on Facebook it has analytics and advertising options not available to individuals or groups.

Business Twitter Account and Personal Twitter Account:

Twitter is important for the demographic it reaches and for staying on top of trends. It isn't something to invest heavily in necessarily, but it is easy enough to perfect at this stage, set up processes, and continue to the next items on the list. Always pursue joy and positivity on Twitter because text-based messages lack all the verbal cues which calm readers. Basically, the more text and the less videos or images, then the more anger that is possible. Tread carefully.

LinkedIn Account and LinkedIn Page:

LinkedIn is crucial for reputability and connecting with professionals—especially e-volunteers. Look up how to do this right and focus on your needs for e-volunteers to really maximize the unique social network on LinkedIn.

Google+ Account and Google+ Page:

Ranking high in Google's algorithms is key for traffic and success. Digital altruism is so easy because so many people have access to software applications, computers, and the Internet. However, when digital creation is easy, the online market is flooded with more supply than demand. This is where having an actual free-standing, physical nonprofit organization or business that is in a building, not solely online, is very helpful. If you don't have a physical organizational presence besides a home office or a common-use facility, then you might want to postpone utilizing Google+ but not for long. You don't have to have a physical place of business, but it helps with SEO.

Business Instagram Account and Personal Instagram Account:

Instagram is so low because most businesses and agencies are still learning how to benefit from Instagram as much as they benefit from the other social media for making sales. However, if you have a lot of products, great customer stories, or purely focus on awareness, then Instagram might need to be a higher priority on this list for you.

Facebook Public Figure Page for Director or Founder:

A public figure page on Facebook acts like another business page and does not get the same tools and organic reach available to a personal Facebook account. This is why many people start a secondary personal Facebook account which Facebook discourages but for the most part allows if this "second you" is significantly different than your primary account. Many people go this second Facebook personal account route with their "business" or "mission" side to show that this is who they truly are or their main mission in life. They then use the secondary Facebook profile more like a LinkedIn profile within Facebook. Eventually they turn it into an official Facebook public figure page. Most people get their friend list on their secondary Facebook account up to 5,000 and then convert to a public figure account at some point afterward.

Appendix D
E-volunteering Case Study: Hurricane Harvey

Hurricane Harvey cost the U.S. Gulf Coast and Houston $180 billion. That's 180,000,000,000 dollars. It's almost unfathomable, and it was the worst economic disaster ever recorded in the United States.

What can one individual do? Such a large number can make you feel alone in the tragedy. Fortunately, the response from Americans near and far—and next door—was unprecedented. Houston—known for being the most diverse city in the south—was a model of togetherness, love, rescue, and problem solving.

Network news kept showing people trapped on their roofs surrounded by water. I watched from Fort Worth and helped share via social media to get boats and rescuers to those desperate Houstonians. 911 was overwhelmed with callers needing rescue from the flood waters. When the authorities needed help, Americans volunteered in massive numbers in person and online. It

E-volunteer Case Study: Hurricane Harvey

was incredible to see Americans working together online and in person to coordinate rescue, relief efforts, jet skis, and recreational boats. It was digital altruism. It was e-volunteering.

Even when the rain stopped, the flood waters were still rising. In 24 hours, my wife—a Houston native—gathered from friends in Fort Worth 1,000 pounds of the non-perishables, diapers, trash bags, and other items requested by disaster relief centers. I did some e-volunteering and documented it on social media—not much really.

I wanted to do more. I had two daughters born with severe disabilities. I knew the loneliness of tragedy. So I set out to gather the nicest $200 basketball shoes to send to victims in Houston to remind them, "We're here now, and we'll be here long term!" From my experiences with tragedy, I knew that a huge, shocking gift can do wonders. A big commitment of self-sacrificing grace and mercy at the beginning can be worth ten times more than it cost the giver. Love and altruism makes a difference.

When I called relief coordinators to ask who needed these huge gifts of encouragement, here's what they told me:

"These people are going to lose their homes!"

Wait. What? That doesn't make a lot of sense. They lost all their possessions...

"The people," the coordinators repeated, "have homes filled up with two feet of mud and soaked to the studs."

"What can we do?" I asked.

"We have more than 100,000 flooded homes," came the reply. "If we don't get them cleaned out and dried out in three weeks then all the homes will be torn down."

E-volunteer Revolution

He continued with the specific needs: "We need volunteers to help get the mud out, pull out all the furniture, tear out the carpet and baseboards, rip off the dry wall, and dry these homes out to prevent mold and decay."

Wow. I can help with this. I've previously built the exact software that solves these problems.

Over the next 48 hours, I worked as an e-volunteer in Fort Worth to help disaster relief coordinators in Houston. Over 100,000 homeowners needed help to save their homes. They had three weeks. We had three weeks.

We had to organize the teams mudding out the flooded homes. Efficiency was crucial. You can't mud out 100,000 homes in three weeks without efficiency. We had to create an app. We made a list of what the relief coordinators needed, and I rapidly made and remade online prototypes.

We solved the biggest problems. Prior to this, teams of men with heavy tools had shown up at homes just to find the homes locked, the owners absent, and cell phones unanswered. Do you break down a door to save someone's home? Do you risk the liability? Do you sit around for a couple hours? People were losing their soaked homes due to inefficiencies which technology could eliminate.

We made intake smooth by making it online not over the phone. We repurposed an online marketplace website into a relief coordination app. Now we had the ability to map, track, and chat. We even let people sign in with their social media profiles! We solved how many people to send to each house by sneaking in questions about the size of the affected home. We prioritized homeowners who put their cell phone number and filled

E-volunteer Case Study: Hurricane Harvey

out which time of the day they would be home. We also provided safety instructions and let homeowners list where they were in the six step mudding out process.

In the first few hours we got 30,000 views. The homeowners started pouring in, and I was shocked—awoken—at the specific demographic asking for help to save their homes!

Who do you think reached out for help? It was the people who needed help—who felt alone. The elderly, the disabled, the single moms. It seemed to me like all the people who were not getting enough support from the community were reaching out to us. Awesome! That's who needs the help.

I really felt like I would have been in their shoes—feeling alone, feeling worried. I have two disabled daughters and I can't always be as involved as I'd like to be in the community. If a flood struck my neighborhood, I might be near the end of the list to receive help because I haven't been there for others as much as I'd like to be—I've got therapies and doctors' visits for my kids. I might be like them and not have a long list of friends to call for help. So when an endless list of moms, grandmothers, and guys in wheelchairs asked for help—it made me cry. Makes me cry now too. Sort of like tears of—not joy—but of being overwhelmed. Overwhelmed with people's willingness to be vulnerable and overwhelmed with the knowledge that they would get help. They were not alone anymore. I'd call those tears of grace.

So that is e-volunteering on my part helping to power the boots on the ground. My part would not have existed without the Houstonians and Americans working nonstop to save people's homes.

The volunteers used the app that they designed

to save at least $15,000,000 of homes from being destroyed by black mold and decay. I was able to help e-volunteer from Fort Worth because I had spent the past year building online marketplaces. When they told me what they needed, I had the exact skillset to meet their need. My e-volunteering was online from start to finish. I held my daughter in my arms while I created the prototype on my computer.

This was a match of our common missions and it was a match of my specific skillset with the relief workers' need. I don't expect every 48 hours of effort to save $15,000,000 in homes of our hard-working neighbors, but I do know the joy of volunteering and e-volunteering. I do want others to have the joy of volunteering and e-volunteering. Digital altruism is powerful. The e-volunteer revolution is here.

About the Author

As a dad of two extraordinary daughters who each astound me with their curiosity, joy, and selflessness—one on her prosthetic leg and the other in her wheelchair—I expect big things from e-volunteers and digital altruism. Instead of in garages and basements, we will see new businesses and movements start in soup kitchens, in Haiti, in speech therapy rooms, and in hospital rooms.

-D.L. Frugé

About Red Badge

Mission Statement

Red Badge was born in 2016 so that no one would die with their "music" still inside them. Red Badge focuses on publishing the experiences, understanding, and views of the disenfranchised, the oppressed, the sick, the disabled, and the failures. Red Badge does not meet people "halfway" but instead meets them wherever necessary to keep that music from dying within them.

Red Badge depends on e-volunteers and donations to assist authors in sharing their experiences, understanding, and views. From start to finish and through multiple editions the process involves the help of e-volunteers.

www.ingramcontent.com/pod-product-compliance
Lightning Source LLC
Chambersburg PA
CBHW031415210526
45464CB00005B/1897